TRANSIENT STABILITY OF POWER SYSTEMS
A Unified Approach to Assessment and Control

THE KLUWER INTERNATIONAL SERIES IN ENGINEERING AND COMPUTER SCIENCE

Power Electronics and Power Systems
Series Editor
M. A. Pai

Other books in the series:

MAINTENANCE SCHEDULING IN RESTRUCTURED POWER SYSTEMS
　M. Shahidehpour and M. Marwali, ISBN: 0-7923-7872-5
POWER SYSTEM OSCILLATIONS
　Graham Rogers, ISBN: 0-7923-7712-5
STATE ESTIMATION IN ELECTRIC POWER SYSTEMS: *A Generalized Approach*
　A. Monticelli, ISBN: 0-7923-8519-5
COMPUTATIONAL AUCTION MECHANISMS FOR RESTRUCTURED POWER INDUSTRY OPERATIONS
　Gerald B. Sheblé, ISBN: 0-7923-8475-X
ANALYSIS OF SUBSYNCHRONOUS RESONANCE IN POWER SYSTEMS
　K.R. Padiyar, ISBN: 0-7923-8319-2
POWER SYSTEMS RESTRUCTURING: *Engineering and Economics*
　Marija Ilic, Francisco Galiana, and Lester Fink, ISBN: 0-7923-8163-7
CRYOGENIC OPERATION OF SILICON POWER DEVICES
　Ranbir Singh and B. Jayant Baliga, ISBN: 0-7923-8157-2
VOLTAGE STABILITY OF ELECTRIC POWER SYSTEMS, Thierry
　Van Cutsem and Costas Vournas, ISBN: 0-7923-8139-4
AUTOMATIC LEARNING TECHNIQUES IN POWER SYSTEMS, Louis A.
　Wehenkel, ISBN: 0-7923-8068-1
ENERGY FUNCTION ANALYSIS FOR POWER SYSTEM STABILITY,
　M. A. Pai, ISBN: 0-7923-9035-0
ELECTROMAGNETIC MODELLING OF POWER ELECTRONIC
　CONVERTERS, J. A. Ferreira, ISBN: 0-7923-9034-2
MODERN POWER SYSTEMS CONTROL AND OPERATION, A. S. Debs,
　ISBN: 0-89838-265-3
RELIABILITY ASSESSMENT OF LARGE ELECTRIC POWER SYSTEMS,
　R. Billington, R. N. Allan, ISBN: 0-89838-266-1
SPOT PRICING OF ELECTRICITY, F. C. Schweppe, M. C. Caramanis, R. D.
　Tabors, R. E. Bohn, ISBN: 0-89838-260-2
INDUSTRIAL ENERGY MANAGEMENT: *Principles and Applications*,
　Giovanni Petrecca, ISBN: 0-7923-9305-8
THE FIELD ORIENTATION PRINCIPLE IN CONTROL OF INDUCTION
　MOTORS, Andrzej M. Trzynadlowski, ISBN: 0-7923-9420-8
FINITE ELEMENT ANALYSIS OF ELECTRICAL MACHINES, S. J. Salon,
　ISBN: 0-7923-9594-8

TRANSIENT STABILITY OF POWER SYSTEMS

A Unified Approach to Assessment and Control

Mania PAVELLA
University of Liège, Belgium

Damien ERNST
University of Liège, Belgium
Research Fellow, FNRS

Daniel RUIZ-VEGA
University of Liège, Belgium

Kluwer Academic Publishers
Boston/Dordrecht/London

Distributors for North, Central and South America:
Kluwer Academic Publishers
101 Philip Drive
Assinippi Park
Norwell, Massachusetts 02061 USA
Telephone (781) 871-6600
Fax (781) 871-6528
E-Mail <kluwer@wkap.com>

Distributors for all other countries:
Kluwer Academic Publishers Group
Distribution Centre
Post Office Box 322
3300 AH Dordrecht, THE NETHERLANDS
Telephone 31 78 6392 392
Fax 31 78 6546 474
E-Mail services@wkap.nl>

 Electronic Services <http://www.wkap.nl>

Library of Congress Cataloging-in-Publication

Pavella, Mania, 1934-
   Transient stability of power systems : a unified approach to assessment and control / Mania Pavella, Damien Ernst, Daniel Ruiz-Vega.
      p. cm. -- (Kluwer international series in engineering and computer science ; SECS 581)
   Includes bibliographical references and index.
   ISBN 0-7923-7963-2 (alk. paper)
   1. Electric power systems--Electric losses. 2. Transients (Electricity) 3. Electric power systems--Control. 4. Electric power systems--Protection. I. Ernst, Damien, 1975- II. Ruiz-Vega, Daniel, 1968- III. Title. IV. Series.

TK1010 .P38 2000
621.31'2--dc21

00-061052

Copyright © 2000 by Kluwer Academic Publishers.

All rights reserved. No part of this publication may be reproduced, stored in a retrieval system or transmitted in any form or by any means, mechanical, photo-copying, recording, or otherwise, without the prior written permission of the publisher, Kluwer Academic Publishers, 101 Philip Drive, Assinippi Park, Norwell, Massachusetts 02061

Printed on acid-free paper.

Printed in the United States of America

Contents

Preface	xiii
Notation	xvii

1. CHAPTER 1 - BACKGROUND ... 1
    1. INTRODUCTION ... 1
    2. SECURITY: DEFINITIONS AND STUDY CONTEXTS ... 3
        2.1 Definitions and classification ... 3
        2.2 Operating modes ... 5
        2.3 Preventive TSA&C. Corresponding needs ... 6
            2.3.1 Power system planning ... 7
            2.3.2 Operation planning ... 7
            2.3.3 Real-time operation ... 8
        2.4 Emergency mode ... 8
        2.5 Security in a liberalized environment ... 9
            2.5.1 Restructured power systems: an introduction ... 9
            2.5.2 Congestion management and ATC ... 10
            2.5.3 OPF: a comeback ... 12
    3. MODELS ... 13
        3.1 General modeling ... 13
        3.2 Static and dynamic models ... 13
        3.3 Transient stability models ... 14
    4. TRANSIENT STABILITY: TIME-DOMAIN APPROACH ... 15
    5. DIRECT APPROACHES - AN OVERVIEW ... 17
        5.1 Brief introductory notice ... 17
        5.2 Application of direct methods to transient stability ... 18
            5.2.1 Introduction ... 18
            5.2.2 Principle ... 19
            5.2.3 Discussion ... 20
        5.3 Past and present status of direct approaches ... 20
            5.3.1 Anticipated advantages and difficulties met ... 20
            5.3.2 The two families of hybrid solutions ... 21
            5.3.3 Concluding remarks ... 23
    6. AUTOMATIC LEARNING APPROACHES - A DIGEST ... 23

		6.1	Problem statement	24
		6.2	Overview of AL methods	24
			6.2.1 Decision trees (DTs)	25
			6.2.2 Artificial neural networks	27
			6.2.3 Statistical pattern recognition	28
			6.2.4 Hybrid AL approaches	28
		6.3	Performances and assets	29
			6.3.1 Overall comparison	29
			6.3.2 Main assets of AL methods	30
		6.4	Comparison of methods	31
	7.	SCOPE OF THE BOOK		31
2.	CHAPTER 2 - INTRODUCTION TO SIME			33
	1.	FOUNDATIONS		33
		1.1	OMIB: concept and variants	33
			1.1.1 Time-invariant OMIB	34
			1.1.2 Time-varying and generalized OMIBs	34
		1.2	From EEAC to SIME	35
		1.3	Principle	37
	2.	GENERAL FORMULATION		37
		2.1	Critical machines identification	38
		2.2	Derivation of OMIB time-varying parameters	39
		2.3	Equal-area criterion revisited	41
		2.4	Stability conditions	43
			2.4.1 Conditions of unstable OMIB trajectory	43
			2.4.2 Conditions of stable OMIB trajectory	44
			2.4.3 Borderline conditions of OMIB trajectory	44
			2.4.4 Objectivity of the stability criteria	45
	3.	STABILITY MARGINS		45
		3.1	Unstable margin	45
		3.2	Stable margin	46
			3.2.1 Remark	46
			3.2.2 Triangle approximation	47
			3.2.3 Weighted least-squares approximation	48
			3.2.4 Triangle vs WLS approximation	48
			3.2.5 Note on sensitivity analysis by SIME	48
		3.3	Existence and range of stability margins	49
			3.3.1 General description	49
			3.3.2 Illustrations	50
			3.3.3 Variation of salient parameters with t_e	50
		3.4	A convenient substitute for unstable margins	52
		3.5	Next candidate CMs and margins	53
		3.6	Normalized margins	55
	4.	SIME's TYPICAL REPRESENTATIONS		55
		4.1	Illustrations on the three-machine system	55
			4.1.1 Stability conditions	55
			4.1.2 OMIB parameters and numerical results	56
			4.1.3 SIME's three representations	56
		4.2	Illustrations on the Hydro-Québec system	58

				Contents	vii

		4.3	SIME as a reduction technique	61
	5.	BACK- AND MULTI-SWING PHENOMENA		61
		5.1	Definitions	61
		5.2	Analytical expression of margins	62
	6.	DIRECT PRODUCTS AND MAIN BY-PRODUCTS		62
		6.1	Description	62
		6.2	Organization of topics	63
	7.	PREVENTIVE vs EMERGENCY SIME		63
		7.1	Preventive transient stability assessment	64
		7.2	Predictive transient stability assessment	64
		7.3	Control	64
3.	CHAPTER 3 - SENSITIVITY ANALYSIS			67
	1.	ELEMENTS OF SENSITIVITY ANALYSIS		68
		1.1	Problem statement	68
		1.2	Sensitivity analysis of the linearized system	69
		1.3	Sensitivity analysis of the supplementary motion	70
		1.4	Synthetic sensitivity functions (ssfs)	71
		1.5	Illustrative examples	72
			1.5.1 Simulation conditions	72
			1.5.2 Discussion	73
			1.5.3 Supplementary motion of state variables	75
			1.5.4 Supplementary motion of time-varying ssf	76
			1.5.5 Discussion	77
		1.6	Supplementary motion of time-invariant ssfs	78
	2.	SIME-BASED SENSITIVITY ANALYSIS		80
		2.1	Specifics and scopes	80
		2.2	On the validity of linearized approximations	81
	3.	COMPENSATION SCHEMES (CSs)		83
		3.1	General scope and principle	83
		3.2	CSs appraising critical clearing times	83
			3.2.1 Description	83
			3.2.2 Illustrative examples	85
		3.3	CS appraising power limits	87
			3.3.1 Principle	87
			3.3.2 Power limit of OMIB	87
			3.3.3 Power limits of system machines	88
			3.3.4 Discussion	89
			3.3.5 Numerical example	89
			3.3.6 Computing areas of the CS	90
4.	CHAPTER 4 - PREVENTIVE ANALYSIS AND CONTROL			93
	1.	PRELIMINARIES		93
		1.1	Chapter overview	93
		1.2	A measure for assessing computing performances	95
	2.	STABILITY LIMITS		95
		2.1	Basic concepts	95
		2.2	Critical clearing times	96
			2.2.1 Basic procedure	96

		2.2.2	Parameters and technicalities	97
		2.2.3	Initial clearing time conditions	98
		2.2.4	Performances	98
		2.2.5	Illustrations on the 3-machine system	100
		2.2.6	Illustrations on the 627-machine system	101
	2.3	Power limits	102
		2.3.1	Preliminaries	102
		2.3.2	"Pragmatic" approach	104
		2.3.3	SIME-based approaches	104
		2.3.4	Variants of the SIME-based approach	106
		2.3.5	Discussion	106
		2.3.6	Performances	107
		2.3.7	Illustration of SIME-based computations	107
		2.3.8	Observations and comparisons	109
		2.3.9	"Pragmatic" vs SIME-based stabilization	111
		2.3.10	Concluding remarks	112
	2.4	Stability limits approximate assessment	112
		2.4.1	Scope and principle	112
		2.4.2	Two-margin approximation	113
		2.4.3	Single-margin approximation	114
		2.4.4	Stopping criteria for first-swing screening	114
		2.4.5	Illustrations on the 3-machine system	114
3.	FILTRA		115
	3.1	Scope of contingency filtering, ranking, assessment	115
	3.2	Basic concepts and definitions	116
	3.3	General design	118
		3.3.1	Contingency filtering block	118
		3.3.2	Contingency ranking and assessment block	118
		3.3.3	Remarks	119
	3.4	A particular realization of FILTRA	119
		3.4.1	Contingency filtering	119
		3.4.2	Contingency ranking	120
		3.4.3	Refined ranking of harmful contingencies	120
		3.4.4	Assessment of harmful contingencies	121
		3.4.5	Computing requirements of FILTRA	121
		3.4.6	Main properties of FILTRA	121
	3.5	Illustrating FILTRA techniques	122
		3.5.1	Simulation description	122
		3.5.2	Zooming in harmful contingencies	122
		3.5.3	Performances	124
		3.5.4	Zooming in classification ability of FILTRA	125
	3.6	Variants of the filtering block	125
	3.7	Concluding remarks	127
4.	PREVENTIVE CONTROL	128
	4.1	Generalities	128
	4.2	Single contingency stabilization	128
		4.2.1	Principle of generation reallocation	128
		4.2.2	Illustration	129
		4.2.3	Comparing stabilization patterns	130

		4.3	Multi-contingency simultaneous stabilization	132
			4.3.1 Principle of generation reallocation	132
			4.3.2 Illustration on the 3-machine system	132
			4.3.3 Illustration on the Brazilian system	133
			4.3.4 Stabilizing inter-area mode oscillations	134
5.	CHAPTER 5 - INTEGRATED TSA&C SOFTWARE			139
	1.	INTEGRATED SOFTWARE		139
		1.1	Basic TSA&C software	139
		1.2	Multi-objective TSA&C software	141
		1.3	Adapting the basic OPF algorithm	143
	2.	A CASE-STUDY		144
		2.1	Maximum allowable transfer: problem statement	144
		2.2	Plant mode instability constraints	145
			2.2.1 Problem description	145
			2.2.2 Contingency filtering, ranking, assessment	146
			2.2.3 Contingencies' simultaneous stabilization	149
			2.2.4 Logical rule vs OPF-based procedures	152
		2.3	Inter-area mode instability constraints	157
			2.3.1 Problem description	157
			2.3.2 Base case conditions	157
			2.3.3 Contingency filtering, ranking, assessment	158
			2.3.4 Contingencies' simultaneous stabilization	159
			2.3.5 NMs' generation rescheduling via OPF	161
		2.4	Concluding remarks	164
	3.	TSA&C IN CONTROL CENTERS		165
		3.1	Introduction	165
		3.2	TSA&C in the EMS	166
			3.2.1 On-line TSA&C for the TSP	166
			3.2.2 On-line TSA&C for the ISO	167
		3.3	Congestion management	168
		3.4	TSA&C for the DTS and Study Environments	169
6.	CHAPTER 6 - CLOSED-LOOP EMERGENCY CONTROL			171
	1.	OUTLINE OF THE METHOD		172
		1.1	Definitions	172
		1.2	Scope	172
		1.3	Principle	173
		1.4	General organization	173
		1.5	Computational issues	174
			1.5.1 Involved tasks	175
			1.5.2 Corresponding durations	175
		1.6	Notation specific to Emergency SIME	175
	2.	PREDICTIVE SIME		176
		2.1	Description	176
		2.2	Procedure	176
		2.3	Remark	178
		2.4	Specifics	178
		2.5	Salient features	179

	3.	EMERGENCY CONTROL	179
		3.1	General principle	179
		3.2	Generation shedding	180
			3.2.1 Computing stability margins	180
			3.2.2 Identification of the machine(s) to shed	182
			3.2.3 Straightforward improvements	183
	4.	SIMULATIONS	183
		4.1	Description	183
		4.2	Simulation results of Predictive SIME	184
		4.3	Simulation results of Emergency Control	185
	5.	DISCUSSION	186
		5.1	Summary of method's features	186
		5.2	Topics for further research work	187
7.	CHAPTER 7 - RETROSPECT AND PROSPECT	189
	1.	SIME: HINDSIGHT AND FORESIGHT	189
		1.1	SIME: a unified comprehensive approach	189
		1.2	Preventive SIME	190
		1.3	Emergency SIME	192
		1.4	Preventive vs emergency control	193
			1.4.1 Description of physical phenomena	193
			1.4.2 Controlled generation vs control time	194
			1.4.3 Discussion	194
		1.5	General comparisons	196
			1.5.1 Description	196
			1.5.2 Preventive vs emergency control	197
			1.5.3 Open-loop vs closed-loop EC	197
	2.	AN ILLUSTRATION	198
		2.1	Description	198
		2.2	Application	199
		2.3	Simulations results	199
	3.	COMPARING CLASSES OF METHODS	201
		3.1	Practical aspects of AL approaches	202
			3.1.1 Real-world applicability concerns	202
			3.1.2 SIME as compared with AL approaches	202
		3.2	Synthetic comparison	203
		3.3	The criteria	203
		3.4	Comments	205
Appendices	207
A– THE EQUAL-AREA CRITERION	207
	1.	GENERAL CONCEPTS	207
		1.1	Introduction	207
		1.2	Principle	208
		1.3	Two-machine system	211
	2.	APPLICATION EXAMPLE	211
B– DATA OF SIMULATED SYSTEMS	215
	1.	THREE-MACHINE TEST SYSTEM	215

2.	HYDRO-QUEBEC POWER SYSTEM	217
3.	EPRI AMERICAN TEST SYSTEMS	218
	3.1　Test power system C	219
	3.2　Test power system A	219
4.	BRAZILIAN POWER SYSTEM	221

References 223
Index 235

Preface

The market liberalization is expected to affect drastically the operation of power systems, which under economical pressure and increasing amount of transactions are being operated much closer to their limits than previously. These changes put the system operators faced with rather different and much more problematic scenarios than in the past. They have now to calculate available transfer capabilities and manage congestion problems in a near on-line environment, while operating the transmission system under extremely stressed conditions. This requires highly reliable and efficient software aids, which today are non-existent, or not yet in use.

One of the most problematic issues, very much needed but not yet encountered today, is on-line dynamic security assessment and control, enabling the power system to withstand unexpected contingencies without experiencing voltage or transient instabilities. This monograph is devoted to a unified approach to transient stability assessment and control, called SIngle Machine Equivalent (SIME).

SIME is a hybrid direct-temporal method which processes information about the system behavior in order to get one-shot stability assessment in the same way as direct methods. It crystalizes a long research effort developed by the group at University of Liège in the field of direct approaches to transient stability. Two distinct methodologies have thus emerged over the years, depending upon the type of temporal information used. The one, called "preventive SIME", relies on time-domain (T-D) programs to get information about simulated stability scenarios of *anticipated contingencies*. The second, called "emergency SIME", uses real-time measurements which take into account the *actual occurrence of a contingency*.

The preventive SIME combines advantages of T-D and of direct methods, evades their difficulties and broadens their capabilities. In particular, it preserves the assets of T-D methods: accuracy, and ability to handle any power system modeling, to process any contingency scenario and to analyze any type

xiv   TRANSIENT STABILITY OF POWER SYSTEMS

of instability (first- or multi-swing; plant or inter-area mode oscillations). Besides, it speeds up the assessment of T-D methods, and opens new possibilities so far considered beyond reach. Such an important achievement is control, i.e., design of near optimal preventive countermeasures.

The emergency SIME, on the other hand, processes real-time measurements in order to design, in real time, corrective actions able to contain the system's loss of synchronism, just *after a contingency has actually occurred.* The emergency control is performed in a closed-loop fashion, comprising predictive stability assessment, predictive control design, and predictive assessment of the effectiveness of the triggered control action.

Despite their conceptual differences, both approaches rely on the same principles.

The content of the book is divided in seven chapters and two appendices. The first chapter deals with the transient stability problem, its formulation, modeling, conventional analysis and a historical overview of the two non-conventional classes of methods available today: direct and automatic learning ones. Reviewing direct methods contributes to understanding the incentives having led to the development of hybrid direct ones, like SIME. On the other hand, the overview of automatic learning methods intends to inform about the essentials of this important class of emerging transient stability methods. The chapter ends up with a restatement of the scope of the book.

Chapters 2 and 3 are the core of the general SIME methodology, whatever the subsequent developments and uses.

Chapter 4 deals with the foundations of the preventive SIME and covers all aspects of transient stability assessment and control. In particular, contingency filtering and ranking, contingency assessment, and (simultaneous) stabilization of harmful contingencies, i.e. control.

The various functions developed in Chapter 4 are organized in Chapter 5 in an integrated transient stability assessment and control package. Its interface with an OPF algorithm is also advocated and illustrated on a case-study using a real-world problem. Further, the integration of this package in an energy management system or dispatcher training simulator environment is also considered for real-time operation.

Chapter 6 lays the foundations of the emergency SIME method, designed for real-time closed-loop emergency control. Finally, Chapter 7 summarizes essential features of preventive and emergency SIMEs, discusses various types of control (preventive, open-loop and closed-loop emergency controls) and finally proposes a comparison of SIME with time-domain and automatic learning methods.

Throughout the monograph, several simulations illustrate and emphasize various facets of the method, using two different types of power systems: a simple 3-machine system, to help the reader understand the basic ideas and to allow easy simulations, using for example MATLAB. The second type concerns real-world or realistic large-scale power systems. They contribute to a better understanding of complex transient stability phenomena and illustrate how the method handles them.

The monograph gives a detailed account of the method and provides the material necessary for all those who, one way or another, want to learn and use it. More generally, the monograph is intended for researchers or utility engineers who want to develop various types of transient stability software packages (e.g., contingency filtering and ranking; real-time preventive control; integrated schemes for transient stability-constrained available transfer capability calculations; real-time closed-loop emergency control; etc.). In short, for all those who are involved in software development and implementation of dynamic security tools in planning studies, in control center energy management systems, or in dedicated emergency controls of important power system sites.

Throughout the preparation of the monograph, we have greatly benefited from the help and advice of many collaborators and PhD students. The SIME method itself is based on the PhD thesis of Dr Yiwei Zhang who elaborated a hybrid version of the EEAC method, previously developed by Dr Y. Xue in his PhD thesis. The method has evolved over the years, and has been tested thanks to various collaborations with industries, in particular with Electricité de France, Hydro-Québec, Electrabel and more recently with Electric Power Research Institute. We also acknowledge the Consejo Nacional de Ciencia y Tecnologia, Mexico for supporting the PhD studies of Daniel Ruiz-Vega. Finally we are most grateful to Mrs Marie-Berthe Lecomte, who patiently and skillfully typed the text.

Notation

All abbreviations, acronyms and symbols are fully defined at the place they are first introduced. As a convenience to the reader, we have collected below some of the more frequently used ones in several places.

ABBREVIATIONS AND ACRONYMS

ATC	:	available transfer capability
CCT	:	critical clearing time (of a contingency); denoted also t_c
CT	:	clearing time (of a contingency); denoted also t_e
CM	:	critical machine
EAC	:	equal-area criterion
EMS	:	energy management system
FACTS	:	flexible alternating current transmission system
MIP	:	maximum integration period
NM	:	non-critical machine
OMIB	:	one-machine infinite bus
SIME	:	single machine equivalent
sTDI	:	seconds of T-D integration
SVC	:	static VAR compensator
TSA	:	transient stability assessment
TSA&C	:	transient stability assessment and control
T-D	:	time-domain
TSL	:	transient stability-limited
VAR	:	volt-ampere-reactive
$3\phi SC$	:	three-phase short circuit
t_e		clearing time (of a contingency); denoted also CT

BASIC SYMBOLS

Unless otherwise specified, the following symbols refer to OMIB parameters:

- P_m : OMIB mechanical (active) power
- P_e : OMIB electrical power
- P_a : OMIB accelerating power
- M : OMIB inertia coefficient
- δ : OMIB rotor angle
- δ_r : OMIB "return" angle, i.e., maximum angular derivation of a stable simulation
- δ_u : OMIB unstable angle
- ω : OMIB rotor speed
- t_u : time to (reach) instability conditions, i.e. to reach δ_u
- t_r : "return" time, i.e. time to reach stability conditions
- η : margin defined by SIME and provided by EAC

SUBSCRIPTS

- D and P stand respectively for "during-fault" and "post-fault" (configuration)
- C stands for "critical"; e.g., M_C : inertia coefficient of the group of CMs
- N stands for "non-critical"; e.g. M_N : inertia coefficient of the group of NMs
- e stands for "elimination"; e.g., t_e : (contingency) clearing time
- r stands for "return"; e.g., δ_r : "return angle"; t_r :"return time"
- st stands for "stable"
- u stands for "unstable"

MISCELLANY

- P_C (active) power generation of the group of CMs
- P_{Ci} (active) power generation of the i-th CM
- P_N (active) power generation of the group of NMs
- P_{Nj} (active) power generation of the j-th NM
- d_i electrical "distance", i.e. angular deviation, between the i-th CM and the most advanced NM

Chapter 1

BACKGROUND

The objectives of this Chapter are:

- *to define the transient stability problem in the realm of power system security, its operating modes, application contexts and corresponding needs (Section 2);*
- *to elaborate on transient stability phenomena and modeling (Section 3) and discuss about main strengths and weaknesses of the conventional time-domain approach (Section 4);*
- *to review the "conventional" direct approaches, examine whether and to which extent they are able to cover (some of) the needs in transient stability studies, and introduce the spirit of the method developed in this monograph (Section 5);*
- *to give a summary of the class of automatic learning methods, so as to subsequently enable comparisons between them and the method developed in this monograph (Section 6);*
- *to describe the scope and overall organization of this monograph (Section 7).*

1. INTRODUCTION

Power systems transient stability phenomena are associated with the operation of synchronous machines in parallel, and become important with long-distance heavy power transmissions.

From a physical viewpoint, transient stability may be defined as the ability of a power system to maintain machines' synchronous operation when subjected to large disturbances.

From the system theory viewpoint, power system transient stability is a strongly nonlinear, high-dimensional problem. To assess it accurately, one therefore has to resort to numerical integration methods, referred to as "time-domain" (T-D) methods. Historically, T-D methods started being used before the advent of numerical computers: calculations of very simplified (and hence of reduced dimensionality) versions of the system dynamic equations were carried out manually to compute the machines' "swing curves", i.e. the machines' rotor angle evolution with time [Park and Bancker, 1929].

Another way of tackling transient stability is a graphical method, popularized in the thirties, and called "equal-area criterion" (EAC). This method deals with a one-machine system connected to an "infinite" bus and studies its stability by using the concept of energy, which removes the necessity of plotting swing curves. EAC has been - and still is - considered to be an extraordinarily powerful tool for assessing stability margins and limits, for evaluating the influence of various system parameters, and more generally for providing insight into the very physical transient stability phenomena. Note that the origin of EAC is not well known; rather, reference is often made to books, like [Dahl, 1938, Skilling and Yamakawa, 1940, Kimbark, 1948], which are among the first to use this criterion.

Actually, the EAC energy concept is a particular case of the Lyapunov's general theory yielding the Lyapunov energy-type function applied to a one-machine infinite bus system.

The Western technical literature recognizes Magnusson [Magnusson, 1947] as being the first to use the concept of energy for studying multimachine power systems transient stability. About 10 years later, he was followed by Aylett [Aylett, 1958]. Magnusson and Aylett may therefore be considered as the forerunners of the application of Lyapunov's method to power system stability. The actual application of the Lyapunov's method to power systems appears for the first time in publications of the "Russian school" (e.g., see Gorev [Gorev, 1960], Putilova and Tagirov [Putilova and Tagirov, 1970] and the references therein); they were followed by two American publications: Gless [Gless, 1966] and El-Abiad and Nagappan [El-Abiad and Nagappan, 1966]. These publications mark the beginning of a tremendous research effort - maybe one of the most intensive in technical matters.

Such a fascination for the Lyapunov approach to power system stability may stem from the appeal of both, the theory as such, and the expected promising capabilities of this approach to cover practical stringent needs that T-D approaches cannot meet satisfactorily. Such needs started being felt round the seventies; and the emerging liberalization of the energy sector renders them even more acute and urgent.

Almost at the same time, other types of approaches started emerging. They were initiated in the context of pattern recognition [Dy-Liacco, 1968, Koizumi et al., 1975]; see also a review of methods in [Prabhakara and Heydt, 1987]. Quite soon, however, it became clear that such methods could not be developed satisfactorily because the computer facilities available in those years were insufficient to meet their demanding needs. Thus, despite thoughtful attempts, these methods started becoming effective only two decades later. Actually, they led to the broader and more elaborate class of "automatic learning" methods. Today, the tremendous progress in computer technology together with the availability of bulky data bases for transient stability purposes yield impressive achievements in this field.

This introductory chapter aims at giving a synoptical but complete picture of recent breakthroughs in transient stability assessment and control before focusing on the very subject matter of this monograph, which is the SIngle Machine Equivalent (SIME) method.

2. SECURITY: DEFINITIONS AND STUDY CONTEXTS

2.1 Definitions and classification

Power system *security* in general may be defined as the system robustness to operate in an equilibrium state under normal and perturbed conditions. Power system security covers a wide range of aspects, usually subdivided into "static" and "dynamic" phenomena. Power system *stability* currently refers to the "dynamic" part of security.

Power system stability may be defined broadly as that property of a power system that enables it to remain in a stable equilibrium state under normal operating conditions and to regain an acceptable equilibrium state after being subjected to a disturbance [Kundur, 1994].

Power system stability is a multifaceted problem depending upon a variety of factors, such as: the time span that must be taken into consideration in order to assess stability/instability; the size of the disturbance considered; the physical nature of the resulting instability. Hence, although a rigorous classification among distinct types of stabilities is difficult, a practical classification often accepted relies on the above factors. Thus, with reference to the time span of the phenomena, one distinguishes short-term from long-term stability. With reference to the size of the disturbance considered, one distinguishes small-disturbance from large-disturbance stability: the former may be handled via linearization of the dynamic equations of motion, while the latter requires nonlinear approaches. Further, both the small-disturbance and large-disturbance stability phenomena may be subdivided into *"voltage"* and *"angle"* ones. Figure 1.1 illustrates this classification. One can see that the angle large-disturbance stability is the so-called *transient stability*; this is one of the two aspects making up what is called *dynamic security*; the other aspect is large disturbance voltage stability.

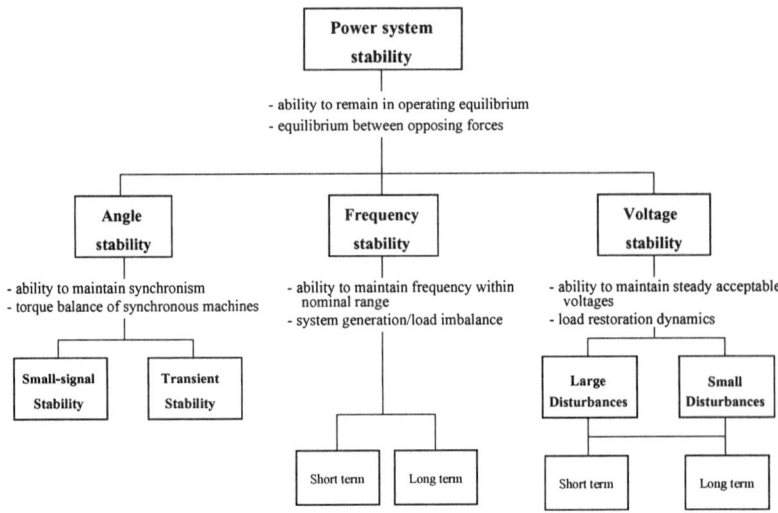

Figure 1.1. Types of power system stability phenomena. Adapted from [Kundur and Morisson, 1997a]

This monograph is devoted to transient stability assessment and control (TSA&C).

Transient stability of a power system is its ability to maintain synchronous operation of the machines when subjected to a large disturbance. The occurrence of such a disturbance may result in large excursions of the system machine rotor angles and, whenever corrective actions fail, loss of synchronism results among machines. Generally, the loss of synchronism develops in very few seconds after the disturbance inception; actually, among the phenomena considered in Figure 1.1 transient stability is the fastest to develop.[1]

The nonlinear character of transient stability, its fast evolution and its disastrous practical implications make it one of the most important and at the same time most problematic issues to assess and even more to control, especially today, with the emerging deregulation practices of the electric sector in many countries.

Indeed, the deregulated electric energy sector in the United States, in Europe, and in many other parts of the world will call for independent system operators to be responsible for the transmission network. The electric utility industry is thus moving into a new regulatory regime, where the new control centers will have to monitor and control networks significantly larger than the existing

---

[1] Large-disturbance voltage stability is another challenging issue of great practical importance, though generally slower to develop than transient (in)stability (typically some (tens of) minutes). In some cases transient voltage instabilities caused mainly by induction motor loads develop as fast as transient (angle) instabilities and the distinction between them is difficult.

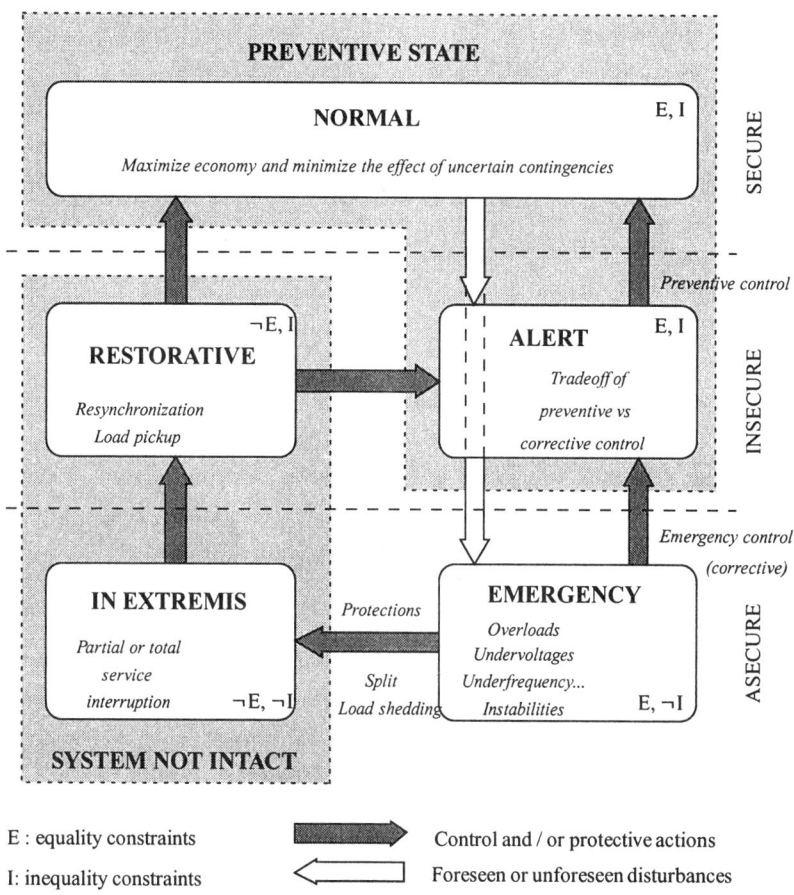

Figure 1.2. Dy Liacco's diagram. Adapted from [Fink and Carlsen, 1978]

ones and to track in much shorter time horizons many more energy transactions than today: larger network size, more important amounts of power flows, and shorter time horizons are likely to impact on TSA&C and to make it even more intricate and indispensable than in the past (see below, § 2.5).

2.2 Operating modes

Most authors credit Dy Liacco for laying down the conceptual foundations of power system security and for defining the different operating modes [Dy-Liacco, 1968]. Figure 1.2 shows a more detailed description of the "Dy-Liacco diagram".

Preventive security assessment is concerned with the question whether a system in its normal state is able to withstand every plausible contingency; if not, preventive control would consist of moving this system state into a secure operating region. However, since predicting future disturbances [2] is difficult, preventive security assessment will essentially aim at balancing the reduction of the *probability* of losing integrity with the economic cost of operation.

Emergency state detection aims at assessing whether the system is in the process of losing integrity, following an actual disturbance inception. In this rather deterministic evolution, response time is critical while economic considerations become temporarily secondary. Emergency control aims at taking fast last resort actions, to avoid partial or complete service interruption.

Restorative mode. When both preventive and emergency controls have failed to bring system parameters back within their constraints, automatic local protective devices will act so as to preserve power system components operating under unacceptable conditions from undergoing irrevocable damages. This leads to further disturbances, which may result in system splitting and partial or complete blackouts. Consequently, the system enters the restorative mode, where the task of the operator is to minimize the amount of un-delivered energy by re-synchronizing lost generation as soon as possible and picking up the disconnected load, in order of priority.

Restorative mode is beyond the scope of this book; rather, direct focus is placed on preventive and emergency TSA&C.

Note. The vertically integrated organization of the electric industry is undergoing a major unbundling and subsequent restructuring into generation, transmission and distribution companies. Nevertheless, the distinction between preventive and emergency operating modes remains unchanged. Further, within the preventive mode, the conventional distinction between planning, operation planning and real-time operation still seems to hold valid, although with some changes. Paragraph 2.3 addresses issues relative to the conventional, integrated type of organization, while § 2.4 discusses the other way of handling TSA&C viz., the emergency mode. Finally, § 2.5 considers changes (anticipated or already in application) of the electric industry restructuring.

2.3 Preventive TSA&C. Corresponding needs

The diversity in power system morphology and operation strategies induces various needs. However, despite their specifics, power systems also have

---

[2] The terms "contingency", "disturbance" and "fault" are used here interchangeably. They refer to a large change in the system state, able to drive the system to insecurity. In the context of transient stability, a contingency may for example be a (three-phase or phase-to-ground) short-circuit at the bus of an important generation plant, the loss of an important tie-line, or the loss of an important load.

common needs, relative to the particular field of application. Traditionally, one distinguishes three such fields, each with its own requirements. They are summarized in Table 1.1 and briefly commented below.

Table 1.1. Application contexts and corresponding needs. Adapted from [Pavella and Murthy, 1993]

Application (time allotted)	Key requirements		Key features	Needs
	Speed	Accuracy		
Expansion planning (months to decades)	Desired	Important (less)	Multitudinous cases	• Screening tools • Sensitivity
Operation planning (months to hours)	Critical	Important	Combining economy with security	• Margins • Sensitivity
Real-time Operation (minutes)	Crucial	Desired	Time is crucial	Real-time (*) • Analysis • Sensitivity • Control

(*) Lack of appropriate techniques may lead to operation stability margins which are needlessly large and anti-economic

2.3.1 Power system planning

In this context, a large number of cases must be performed, months to years before the planned system is finally designed.

Despite the long time allotted to these simulations, speed is yet an important factor. Indeed, to overcome the "curse of dimensionality", a rapid, screening tool is needed to identify the "interesting" situations on which the planner should concentrate. On the other hand, a sensitivity assessment method is desired, able to provide information about the relevant system parameters, their impact on stability, and thence to suggest means to reinforce stability, whenever needed.

2.3.2 Operation planning

In these studies, where the time horizon reduces to days or hours, the speed becomes critical. Moreover, as the power system is operated in ways not necessarily anticipated during the system design[3], there is a need, in addition to the crude stability/instability assessment to appraise stability margins and, whenever necessary, to suggest means to increase them. Once again, sensitivity

---
[3] especially since power systems restructuring, see below, § 2.5.

analysis and means to reinforce stability, in addition to analysis, are essential characteristics needed.

In the present-day practices operation planning stability studies merely reduce to assessment of stability limits or operating guidelines. Techniques able to suggest corrective actions are very much needed.

2.3.3 Real-time operation

Here, only some (tens of) minutes are left to:

- analyze the situation, i.e., screen a large number of contingencies in order to identify the potentially harmful ones,
- scrutinize each one of these latter and design appropriate control actions in case the contingencies do occur,
- decide whether to take actions preventively, or to rely on emergency control, i.e., to postpone actions until such contingencies actually occur.

Hence, speed becomes a crucial factor; besides, control tools are strongly desired, together with sensitivity analyses (inasmuch as they provide early warnings and help determine control actions).

2.4 Emergency mode

As mentioned above, the concern of the preventive mode is to forecast the projected situation, so as to best assess the system stability limits and/or determine appropriate control actions. These may be preventive or emergency control actions (or a combination of both) designed to cope with plausible contingencies identified as dangerous [Kundur and Morisson, 1997b]. The balance between the two options stems from the fact that preventive actions are more versatile and easier to apply than emergency controls, but lead to an increase in operating cost.

However, even in the context of real-time operation, it is very unlikely that such control actions can be optimized in the preventive mode, because of the combinatorial nature of the events which might occur. Ideally, the final control decision should be taken for the *actual* system state, after a (large) disturbance has *actually* occurred. This should be the task of *closed-loop emergency control*.

This type of emergency control is (near)optimum, since it addresses the real problem, but also more difficult to handle, given the extremely short times left to make decisions and take actions able to preserve system's integrity (fractions of second).

The preventive type of approach, on the other hand, generally faces the tradeoff between security and economics in a sub-optimal way. But it is easier to achieve and, so far, it is the only one in use. Yet, today, where new legislations open electricity markets, the trend is to operate the power systems continuously closer to their security limits: the need for emergency control is strongly felt. The question is how to reach this objective.

As will be seen in Section 4, traditional pure analytical T-D approaches are unable to tackle stringent needs imposed by real-time preventive and even more by emergency TSA&C. One thus is led to search for more adequate methods. They have to be found in the realm of "modern" methods, either of the analytical or the automatic learning class (Sections 5 and 6).

2.5 Security in a liberalized environment
2.5.1 Restructured power systems: an introduction

In many countries around the world, electric utilities are being restructured from a traditional vertical model, in which all the activities like generation, transmission and distribution of the electric energy were owned and operated by one single private or state entity, to a horizontal model in which all three activities become independent and, in some cases, are owned by several different companies. This process was performed in order to allow competition between different private companies and utilities, initially at the generation level, and with the trend of extending it to all levels. It was justified mainly by two reasons:

- Prices of electricity in some regions were too high compared to the prices in other regions located in the same country (for example, this happened in USA). In order to allow consumers to pay lower electricity prices, it was thought that increasing competition would make the prices go down to an acceptable level [EIA, 2000].
- In the past, the investment required for building new electric facilities like power plants was so huge that only a very big utility could afford it (economies of scale). This characteristic, which holds still valid in the case of transmission lines, changed recently (ten or fifteen years ago) by the developments achieved in the materials and turbine technology, making the efficiency of smaller power plants comparable to the one of the big thermal plants [Hunt and Shuttleworth, 1996].

Other reasons for unbundling generation from transmission that have been mentioned are: the need of increasing efficiency through better investment decisions and better use of existing plants and the introduction of incentives where conventional utilities have become inefficient. In order to allow competition, almost all the restructured companies were organized in the following way.

Generation, transmission and distribution subsystems were separated and became independent (at least administratively) one from another. In most of the cases, generation and distribution subsystems were privatized or sold after being divided in many parts, in order to foster competition at these levels. Regarding the transmission subsystem, it is universally agreed that it should remain as a monopoly, because of the physical characteristics of power system operation and the advantages of its centralized control. In this case, even if this activity remained as a regulated monopoly, a new figure, the "system operator" was created.

There are many interpretations of the system operator (SO). Its main role is to provide "open, non-discriminatory and comparable" transmission system access to all supplies and loads of electrical energy in their influence areas [Shirmohammadi et al., 1998, Ilic et al., 1998, European Parliament, 1996, FERC, 1996]. In order to achieve this objective, in some market structures it is considered that the SO should be a regulated, but not state owned, non profit entity, completely independent from generation and transmission companies (ex., the California ISO [Alaywan and Allen, 1998]). In other markets, the SO is also the transmission system owner and can be either a for-profit investor owned company, or a state owned one [Dy-Liacco, 1999] (e.g., see UK and for the second France [Russell and Smart, 1997, Merlin et al., 1997]).

Independently of its ownership and although this "system operator" usually has many additional obligations depending on the given new market structure, it is always responsible for power system security assessment and control.

2.5.2 Congestion management and ATC

After deregulation, and additionally to other obligations, the system operator is responsible for managing two main security problems necessary for the new horizontal electricity market structures: congestion management and Available Transfer Capability (ATC) calculations. These two important problems are briefly described below.

Congestion Management. Congestion is a new term (in power systems) that comes from economics. It is being used, after restructuring, for designing situations in which producers and consumers desire to generate and consume electric power in amounts that would cause the transmission system to operate at or beyond one or more of its transfer limits. Congestion management consists of controlling the transmission system so that the transfer limits are observed.

Congestion was present in power systems before de-regulation and was discussed in terms of steady state security. Its basic objective was to control generator output so that the system remain secure (no limits violated), at the lowest cost. Most of the energy sales were between adjacent utilities and a transaction would not go forward unless each utility agreed that it was in their best interest for both economy and security. Problems like the one we call congestion would only arise when the transaction had an impact in the security of a utility not involved in the transaction (third party wheeling) [Christie et al., 2000].

SO creates a set of rules for ensuring sufficient control over producers and consumers in order to maintain acceptable security and reliability levels in both short and long term, while maximizing market efficiency. The rules must be robust because some market participants could exploit congestion to increase profits for themselves at the expense of market efficiency. The rules also should be transparent, fair and clear to all market participants. The form of congestion management is dependent on the structure of the energy market and cannot be separated from market considerations.

Available Transfer Capability. The US Federal Energy Regulatory Commission (FERC) requires, among other things, that the SO (in the US Independent System Operator or ISO) responsible for each regional transmission system monitors and computes the Available Transfer Capability (ATC) for potentially congested transmission paths entering, leaving and inside its network. ATC would be a measure of how much additional electric power (in MW) could be transferred from the starting point to the end point path.

In order to give the same opportunity to all companies owning generation facilities for locating and obtaining transmission services between their generation sites and their customers, while maintaining reliability and security, ATC values for the next hour and for each hour into the futures (next day) is placed in a web site known as the open access same time information system (OASIS), operated also by the ISO. In this way, anyone wishing to send a power transaction on the ISO transmission system would access OASIS web pages and use the ATC information available there to determine if the transmission system could accommodate that transaction, and to reserve the necessary transmission service. [Christie et al., 2000]

New challenges. It is important to mention that only static limits are currently used for both, congestion management and ATC calculations. However, it is generally agreed that all power system security limits (static, voltage and transient) should be taken into account since the limiting condition in some portions of the transmission networks can shift among them over time.

Power system dynamic security assessment and control is now a more demanding task since, in addition to the conventional requirements of operating power systems, it has more constraints, and new obligations and problems like the following ones.

- The new facility for making energy contracts between customers and companies for the long distance power transmission possibly causes third party wheeling. In order to identify and control this congestion problem, it is sometimes necessary that the system operator analyzes larger systems.
- After unbundling the transmission services, many of the services that were naturally provided by the different electric generation plants for controlling

the power system in order to allow system operation (ancillary services) now have to be accounted and paid separately, sometimes in a real-time market that is run and coordinated by the SO [Shirmohammadi et al., 1998, Alaywan and Allen, 1998].

- The efficient utilization of power systems with an acceptable security level is becoming more and more dependent on system controls. Successful energy trading can overwhelm the existing system control structure and the new business structure will have an impact on what controls are used as well as how they are designed and deployed. Since power system physical functioning remains the same, the specification and design of these controls should be part of an overall study carried out by an independent entity (usually the one in charge of security and planning, the SO). Otherwise, system security and economy could be sacrificed, defeating the very purpose of restructuring the industry [Kundur and Morisson, 1998].

- Depending on the structure of the electricity market the old unit commitment process is now replaced by an auction market that sets up the generation profile for the following day or the next hour. The way this auction market works is dependent on the congestion management and/or the ATC calculation; therefore, the market should closely interact with the system operator in order to ensure security assessment and control [Sheblé, 1999].

These demanding needs of the restructured electric markets operation made necessary to use in new ways some tools already available as EMS functions in the control center and to coordinate its execution with some new applications that still need to be developed.

2.5.3 OPF: a comeback

This is the case, for example, of the Optimal Power Flow (OPF). The advent of open access and the competitive market has given OPF a new lease on life and respectability as the indispensable tool for nodal pricing and its variants, zonal and locational pricing [Dy-Liacco, 1999]. Its optimization function also switches from the economical optimization of the system state in order to achieve the minimum operating cost, using the individual cost of the different units, to the optimization for maximizing the profits of the several different players using bids. Its capacity to solve steady state security problems is also used for static security constrained dispatch in congestion management.

Today, there are many situations where stability limits are reached before static constraints like thermal limits. In this kind of stability-limited power systems, it is thus necessary to assess the dynamic security limits of the system in order to provide useful values for ATC and for the congestion management process. These new tools, required for assessing dynamic limits in all the contexts of power system planning and operation, should act in a coordinated way with the existing EMS functions and design applications (like the OPF) already available. This new problem is addressed in Chapter 5 of this monograph, where the combination of a transient stability assessment and control scheme and an OPF is performed in order to provide solutions to power system security problems that satisfy at the same time static and transient stability constraints.

3. MODELS

3.1 General modeling

Dynamic time constants in power systems range from fractions of microseconds (electromagnetic phenomena) to hours (thermal phenomena). As was noted earlier, the time constants of interest in dynamic security assessment correspond to a much smaller window (say, above 1 millisecond and a few minutes). Within this context, power system dynamic behavior is governed by two sets of non-linear equations:

$$\dot{x} = f(x, y, p) \quad (1.1)$$
$$0 = g(x, y, p) . \quad (1.2)$$

Set (1.1) consists of the differential equations (generators, motors, including their controls, and other devices whose dynamics are modeled); set (1.2) comprises the algebraic equations of network and "static" loads. The dimension of vector x depends on the modeling detail; it is lower bounded by twice the number of system machines (e.g., typically ≥ 50), but may be orders of magnitude larger. The dimension of vector y is lower bounded by twice the number of nodes of the power system model (e.g., typically ≥ 1000). Vector p explicitly represents parameters whose influence on dynamic security may be studied (e.g., generator output, load level, interface flows ...) as well as disturbances (represented as a sequence of "instantaneous" changes in p).

Imposing $\dot{x} = 0$ in (1.1) yields the overall equilibrium equations which allow computing the steady state conditions of the system in its pre-fault state or in its post-fault state.

3.2 Static and dynamic models

Assuming that the system is in an acceptable steady-state pre-fault operating regime, the overall TSA problem consists of evaluating the quality of the system response when subjected to various disturbances.[4]

The *static* (part of) security assessment thus consists in evaluating the properties of the post-fault equilibrium state, by checking that it leads to viable and acceptable operating conditions. This implies in particular that the post-

---

[4] Since the system is non-linear, its properties are disturbance dependent.

14   TRANSIENT STABILITY OF POWER SYSTEMS

fault equilibrium is locally stable, and that steady-state voltage magnitudes and currents satisfy a certain number of constraints.

The *dynamic* (part of) security assessment, then considers whether the system would indeed be able to reach its post-fault operating conditions. The dynamics in (1.1) may usually be decomposed into two subsets: one faster, generally only related to the electro-mechanical angle dynamics of the synchronous machines; and one slower, normally only related to the stability of voltage (and load) restoration process due to the action of the slow subset of automatic voltage control devices[5]. They yield tools for respectively transient and voltage stability assessment.

3.3   Transient stability models

In the particular context of transient stability, one may distinguish the following types of state variables (SVs):

- machine state variables
  - the machine per se has a minimum of 2 (mechanical) SVs (rotor angle and speed); in addition, generally 6 (although there could be more) SVs can be considered for a complete electrical model
  - the excitation system of a machine has usually in between 1 and 5 SVs; note, however that consideration of PSSs can increase this number (the output signal of the PSS is sent to the excitation system)
  - turbine and governor of a machine have generally up to 5 SVs (a PSS can be added to the system)
  - all in all, 18 SVs is a standard number of accurate machine modeling;
- load state variables
  - static load: no additional SVs are involved, the characteristic of such a load being a function of bus voltage magnitudes and frequency
  - dynamic load; systems with large number of motors require a dynamic representation involving 1 to 2 SVs per load;
- special devices (SVCs, HVDC links, FACTS, etc): the number of SVs is not standard; it varies with the degree of modeling detail; besides, modeling of such devices generally requires a good number of macroblocks.

More or less detailed description of the above various components of a power system is used in T-D programs to analyze transient stability phenomena.

These phenomena are generally strongly non-linear. They are caused by important disturbances, and characterized by the type of resulting instabilities:

---

[5] Transformer taps, shunt-capacitor switching, secondary voltage control, and overexcitation limiters.

first-swing vs multiswing, upswing vs backswing, plant vs inter-area mode of instabilities, and their combinations.[6]

Conventionally, these phenomena are assessed using T-D programs, as described below. Let us simply mention here that today a rather large number of T-D programs exist, based on more or less detailed models of the power system. They range from general purpose power system dynamic simulation packages, down to simplified models and approaches for the study of a particular subproblem. Below we mention only those which, one way or another, will be used in this monograph: ETMSP [EPRI, 1994]; MATLAB [MATLAB, 1999]; SIMPOW [ABB, 2000]; ST-600 [Valette et al., 1987].

4. TRANSIENT STABILITY: TIME-DOMAIN APPROACH

The conventional time-domain (T-D) approaches assess the system robustness vis-à-vis a given disturbance by solving, step-by-step, eqs (1.1), (1.2) modeled for the transient stability problem, and computing the machine "swing curves" (rotor angle evolution with time), along with other important system parameters.

A disturbance in general is defined as a sequence of events, starting at t_o and finishing up at t_e (the time of its elimination or clearance), when the power system enters its post-fault configuration. Note that this latter configuration depends on the scheme of the disturbance clearance.

Thus, to assess the system robustness vis-à-vis a given disturbance, the T-D approach simulates the system dynamics in the during-fault and post-fault configurations. Generally, the during-fault period is quite short (e.g., 100 ms or so). On the other hand, the post-fault period may be much longer: typically, a system which does not lose synchronism after, say, some seconds, is considered to be stable, i.e., able to withstand the disturbance under consideration. The maximum simulation period depends upon the characteristics of the very power system and the degree of its modeling sophistication. It generally does not exceed 15 s for full detailed modeling, while 3 s are deemed enough for simplified modeling, provided that this latter is valid for assessing first order effects of transient stability phenomena (which may not be the case).

Observe that the definition of criteria detecting the loss of synchronism is also a matter of operational practices; they may differ from one power system to another and from one T-D program to another. They generally depend on maximum deviation of machine rotor angles and rotor speeds. In any case,

---

[6]The various types of instabilities are illustrated on real-world examples throughout the monograph. (For example, Fig. 2.9 illustrates multiswing instabilities, Fig. 4.10 inter-area mode oscillations.) Their various characterizations are summarized in § 1.2 of Chapter 7.

16   *TRANSIENT STABILITY OF POWER SYSTEMS*

these are subjective rather than objective criteria, inspired from the system operator experience.

Typically, T-D methods assess the system robustness vis-à-vis a given disturbance

- either upon fixing a clearing time, t_e, and assessing whether the system loses synchronism or on the contrary remains stable throughout the maximum integration period
- or by assessing stability limits: power limit for a given clearing time, or critical clearing time (CCT) for given pre-fault operating condition; CCT is defined as the maximum duration that a disturbance may remain without the system losing its capability to recover a normal operating condition.

The search of stability limits may, for example, be conducted by dichotomy. In any case, it requires many trials (say, 3 stable and 4 unstable); of them, the stable ones are much more time consuming, since they are pursued for the maximum simulation period. In contrast, the unstable ones are faster (their simulation is generally stopped after some hundred ms, where the system is declared to go out of step); note, however, that this holds true for first-swing instability phenomena: detecting multi-swing instability is more time consuming, since the loss of synchronism may arise after some seconds.

Power limits are more popular in the United States, CCTs in Europe. Actually, computing CCTs is easier than computing power limits, and the usefulness of CCTs goes beyond the mere stability limit computation. For example, they may be used to screen a set of contingencies and rank those found to be "potentially interesting", i.e., those whose CCT is close enough to the operating time of the system protections to clear the considered contingency. CCTs may also be used as the reference for assessing the accuracy of non-conventional methods (direct or automatic learning ones, see below in Sections 5 and 6). Further, for automatic learning methods, CCTs may be used to classify stability scenarios as stable or unstable when constructing the required data base (see also below in Section 6).

The question of concern that now arises is: which are the strengths and weaknesses of T-D methods ? A list is proposed below which yields a twofold, apparently contradictory observation: T-D methods are definitely insufficient, yet indispensable for in-depth transient stability investigations.

Pros. T-D methods are able to:
- provide essential information about relevant parameters of the system dynamic evolution with time (machine swing curves, i.e. rotor angles; speeds; accelerations; powers; etc.);
- consider any power system modeling and stability scenario;
- reach the required accuracy, provided that the modeling of a power system is correctly designed and its parameters accurately known.

Cons. T-D methods are unable to provide:

- straightforward screening tools in order to discard "uninterestingly harmless" disturbances;
- sound stability margins which would inform one about "how far" from (in)stability the system is, and which would yield suitable sensitivity analysis tools;
- guidelines for control.

In short, T-D methods cannot meet major needs identified in Table 1.1, and, in particular, those relating to control, be it of the preventive or the emergency type.

Nevertheless, T-D methods are the reference for transient stability analysis. Besides, they are essential to the design of modern methods as will be seen below: for hybrid direct methods (§ 5.3.2); for automatic learning methods as a tool for assessing stability, i.e., for preanalyzing the cases of the learning set (Section 6).

A final note: the T-D approach has long been considered as very CPU time consuming; it is interesting to observe that within the last years the time required for a single simulation with high order models of a typical power system has shrunk from half an hour to some seconds, essentially thanks to increased CPU speeds of high performance workstations. Today, T-D methods become even "faster than real-time". Yet, the weaknesses identified earlier still hold true.

5. DIRECT APPROACHES - AN OVERVIEW

5.1 Brief introductory notice

The above deficiencies of T-D methods gave an impetus to the development of non-conventional approaches: direct (Lyapunov-like) and automatic learning ones.

Direct methods started being developed in the sixties [Gorev, 1960, El-Abiad and Nagappan, 1966, Gless, 1966, Putilova and Tagirov, 1970]. One main attraction was their capability of restricting the T-D simulations solely to the during-fault period while avoiding all repetitive runs, i.e. of reducing simulations to a very small percentage of the overall computing effort of T-D methods. The other attraction was that direct methods may provide sound stability margins, which in turn open up possibilities towards suitable sensitivity analysis.

The above features of paramount practical importance, together with the fascination of the application of direct methods to power system transient stability, explain the researchers' intensive efforts over almost four decades. Quite soon, however, the first enthusiastic expectations of these methods were overshadowed by two main difficulties. The first is linked to the difficulty of constructing good Lyapunov functions for multimachine power systems, unless (over)simplified modeling is used. The other difficulty is related to the assessment of a practical stability domain, that is suitable from both viewpoints: computational efficiency and accuracy of the transient stability assessment in the large. As often happens in applications, these difficulties have eventually been circumvented by combining theoretically pure approaches with pragmatic engineering solutions. And, as usual, many combinations have been proposed, corresponding to various tradeoffs. They are briefly described below.

5.2 Application of direct methods to transient stability

5.2.1 Introduction

Broadly, the Lyapunov direct method [Lyapunov, 1907] relies on the construction of a Lyapunov V-function, which is a scalar function of the system state vector (x) obeying a number of conditions, in particular: positive-definiteness of V and (semi-)negative-definiteness of its time derivative $\dot{V}$ along the solutions of eqs. (1.1),(1.2). The concept of Lyapunov function is illustrated in Fig. 1.3 for a simple hypothetical "one-machine-infinite-bus" power system, with two state variables δ (machine rotor angle) and $\omega = \dot{\delta}$ (machine rotor speed).

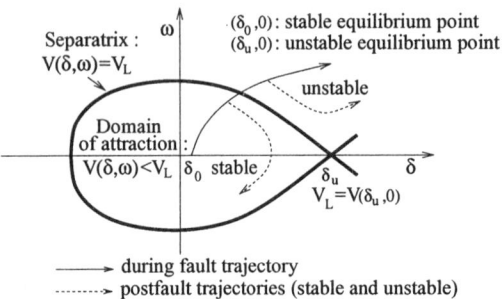

Figure 1.3. Phase plane of "one-machine-infinite-bus" system and exact V-function

The aim of the direct method's application to transient stability was to alleviate some of the deficiencies of the T-D approach.

The following paragraphs deal in a sequence with elements of the application of the Lyapunov direct method to power system transient stability, the two main difficulties met and the ways proposed to circumvent them.

This brief overview will not attempt to make a critical survey nor an exhaustive description of all existing streams. Hence, by no means could it do justice to the impressive number of contributions to the field, which is still being fed by new publications. Rather, it attempts to suggest that, as is often the case in engineering, the combination of good theoretical approaches with physical know-how is able to furnish innovative approaches much more powerful than its components.

5.2.2 Principle

The principle of the application of the Lyapunov criterion to power system transient stability may be stated as follows: under pre-assigned stability conditions (i.e., a pre-assigned disturbance and its clearing scenario) assess whether the system entering its post-fault configuration is stable, using the Lyapunov criterion. In turn, this yields the following practical procedure: upon constructing a Lyapunov function for the post-fault system, compute its value $V(x(t_e))$, where x is computed for the during-fault trajectory, and decide that the system is stable if this is smaller than $V_L = V(x^u)$; it is unstable otherwise. Here, $x(t_e)$ denotes the values taken by the components of the system state vector at t_e, the disturbance clearing time; these values are computed via T-D simulations carried out in the during-fault period for successive values of t_e. On the other hand, x^u denotes the value that the components of the system state vector take at a point located on the boundary of the stability domain, or of its estimate.

$V = V_L$ provides the stability limit condition, i.e. the CCT for a given power level, or the power limit for a given t_e. On the other hand, the excess of V_L over V represents a sound stability margin, measuring "how far from instability" the system is under preassigned conditions. Figure 1.4 portrays schematically the search of a CCT.

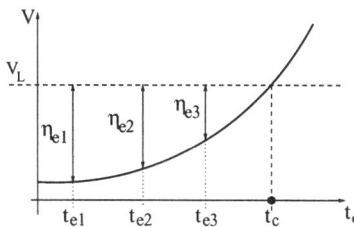

Figure 1.4. Principle of critical clearing time computation by the Lyapunov direct criterion

From a computational point of view, the above procedure requires:

- computation of $V(x(t_e))$ which is almost negligible, provided that $V(x)$ has already been constructed for a given power system model and that $x(t_e)$ is known;

- a single T-D simulation for the during-fault period to compute $x(t_e)$ for successive t_e's; it is upper bounded by the CCT, i.e., it represents a very small percentage of the overall amount of simulation required by the T-D method with equivalent modeling;

20   TRANSIENT STABILITY OF POWER SYSTEMS

- computation of $V_L = V(x^u)$, which is almost negligible for a given x^u; the problem is how to compute that x^u which provides a limit value V_L of practical interest.

5.2.3 Discussion

1. According to the Lyapunov direct method, the study of a power system stability consists of constructing a Lyapunov function for the dynamic equations of motion, and determining the value it takes on the boundary of the stability domain.

Unfortunately, there is no effective method for constructing closed form Lyapunov functions for general dynamical systems. However, for a class of nonlinear systems there are systematic procedures for constructing Lur's type Lyapunov functions with the required sign definite properties. (For details, see [Rozenvasser, 1960, Kalman, 1963, Moore and Anderson, 1968, Willems, 1969, Pai, 1981, Pai, 1989, Pavella and Murthy, 1993]; for a survey see [Ribbens-Pavella and Evans, 1981]).

2. In general, nonlinear systems may have many stable equilibrium points (SEPs); therefore the character of motions close to a SEP does not hold globally for all motions.

3. The Lyapunov stability theorems do not provide conditions for obtaining the largest stability domain estimate (SDE). Rather, they provide sufficient conditions for testing stability via a particular choice of a V−function. But since these conditions are not necessary but sufficient [7], from a practical point of view the concern is to search for a convenient Lyapunov function that ensures a SDE which is as large as possible.

4. To summarize, the choice of a V−function and of its limit value is of paramount importance for making the Lyapunov criterion interesting in transient stability studies, where the concern is to assess accurately stability limits or margins: (over)conservative results that the pure Lyapunov criterion tends to provide (since it guarantees stability but not instability) are of no practical interest.

5.3 Past and present status of direct approaches

5.3.1 Anticipated advantages and difficulties met

The above short description suggests that two main advantages can be expected from the direct approach, namely: (i) straightforward computation of transient stability limits, since direct methods reduce the T-D simulations

---

[7]This is however different from the theorem: "the necessary and sufficient condition for the stability of a dynamic system is that it has a Lyapunov function". This admittedly interesting theorem from a theoretical viewpoint hardly helps construct Lyapunov functions.

solely to the during-fault period while avoiding all repetitive runs; (ii) suitable definition and computation of margins, which in turn allows sensitivity analysis.

At the same time, it appears that to make the method applicable in practice, one should be able to: (i) construct "good" V-functions; (ii) assess "good" limit values or, equivalently, good practical stability domain estimates.

Both of the above conditions have been found to be extremely difficult to meet properly. With reference to condition (i), it appeared quite early that the construction of V-functions was only possible with over-simplified (actually unacceptably simplified) system modeling. Among the many interesting attempts to circumvent this difficulty is the "structure preserving modeling" proposed by [Bergen and Hill, 1981]. Another one, which has finally prevailed, led to the use of "pseudo-Lyapunov functions" consisting of hybridizing V-functions with T-D methods (see below); credit for this may be given to [Athay et al., 1979].

As for condition (ii), among the large variety of solutions proposed are the methods of [Athay et al., 1979], of [Kakimoto et al., 1980], the acceleration approach [Ribbens-Pavella et al., 1981], the BCU or exit point [Chiang et al., 1991], the PEBS [Pai, 1989], and the IPEBS methods [Fonseca and Decker, 1985].

All these methods aim at identifying the value taken by the Lyapunov function at the boundary of the practical stability domain estimate; but they differ in the way of doing so. Broadly, two different approaches may be distinguished; the one consists of computing x^u, the "unstable equilibrium point of concern", and hence $V(x^u)$; the other relies on criteria able to suggest when the system trajectory comes "close enough" to the boundary of the practical stability domain estimate.

Nevertheless, although inventive, the various solutions were not able to properly overcome the difficulties met. Indeed, the proposed stability domain estimates were found to provide overly conservative stability assessments, with an unpredictably varying degree of conservativeness; besides, from a computational point of view, the involved computations were often quite cumbersome, counterbalancing the computer gains expected of direct methods, and even making them much slower than T-D methods.

5.3.2 The two families of hybrid solutions

An impressive research effort and number of publications have been devoted to the derivation of (pseudo)-Lyapunov approaches, able to handle the transient stability problem in a way which is flexible (with respect to power system modeling), accurate (as compared to the corresponding T-D assessment using the same degree of modeling detail) and computationally efficient. They eventually came up (explicitly or implicitly) with the following two observations:

- the stability-domain estimation problem may be tackled by considering a two-machine or a one-machine equivalent of the multimachine power system

- the modeling problem may be solved by hybridizing the direct method with T-D calculations.

The first observation stems from the fact that for the particular case of a two-machine or a one-machine equivalent system described with simplified modeling, the stability condition of the Lyapunov criterion becomes sufficient **and** necessary instead of being solely sufficient. This gave rise on the one hand to the Bellman decomposition-aggregation approach and vector Lyapunov functions, where the multimachine power system is decomposed into 2-machine subsystems [Pai and Narayana, 1975, Grujic et al., 1987][8]; on the other hand, to the single-machine equivalent approaches [Rahimi and Schaffer, 1987, Xue et al., 1988].

The idea underlying the second observation is to construct a Lyapunov function for the simplified power system modeling, $V(x)$, while however computing step-by-step the components of vector x via a T-D program run with the desired detailed modeling. The resulting $V(x)$ becomes path-dependent and is not anymore a true Lyapunov function; nevertheless, from a practical point of view this works nicely with functions of the energy type, like the popular transient energy function (TEF) (e.g., [Athay et al., 1979, Pai, 1989, Fouad and Vittal, 1992, Fonseca and Decker, 1985]). This has led to hybrid approaches, either of the multimachine type (e.g.,[Maria et al., 1990], or of the single-machine equivalent type (e.g., [Zhang et al., 1997a]). The SIME method (for SIngle Machine Equivalent) belongs to this latter family.

The resulting practical approaches are hybrid direct − T-D methods of two types. The one considers a Lyapunov function constructed for the multimachine power system and computed along the multimachine trajectory. The other considers a one-machine equivalent of the multimachine system, and studies its stability using the equal-area criterion (EAC). The SIME method belongs to this latter type.

Careful examination of these two hybrid types suggests that, after all, both of them are capable of providing good practical results. Indeed, they both rely on energy considerations; and the Lyapunov direct method is "just" a generalization of the energy concept.

However, in order for a hybrid function to be interesting in practice, one should be able to assess easily its limit value. The use of EAC simplifies

---

[8] But their subsequent aggregation imposes extremely conservative stability conditions, unacceptable for practical purposes.

greatly this task for the one-machine equivalent, and this factor is shown to be of paramount importance.

5.3.3 Concluding remarks

Power system transient stability phenomena are almost as old as the Lyapunov theory itself [Lyapunov, 1907, Park and Bancker, 1929]. But its application to transient stability started being developed round the middle of the 20th century. The initial incentive was to replace (major part of the) T-D simulations, and thus to alleviate massive numerical computations involved by the intricacy of the very physical problem and by the trend to interconnect power systems thus increasing their size.

Surely, the tremendous fascination for the Lyapunov approach to power system stability may be attributed to its theoretical appeal, and its practical promising outcomes; but curiously enough, to a large extent it is also attributable to the accompanying difficulties at stake. And although the approaches have matured significantly, they still inspire research work.

The dramatic progress in computer technology allowed significant reduction of the computational burden of T-D methods and resulting significant speed up. But at the same time, the difficulties caused by operating the power systems increasingly closer to their security limits bred new needs for ultra-fast analysis, sensitivity analysis and control. Thus, along the years, the incentive for alternative solutions more efficient and powerful than T-D programs has gradually shifted from mere CPU considerations to basic requirements.

The SIME method attempts to meet such requirements.

6. AUTOMATIC LEARNING APPROACHES – A DIGEST

The subject matter of this section is beyond the scope of the monograph. Nevertheless, we thought it interesting to give a brief overview in order to provide a picture of this class of emerging "modern", "non-conventional" methods. This will also allow us to make a comparison of all existing methods at the end of the monograph. The reader interested exclusively in SIME may skip this section.

The review of methods given below is based on work developed by Wehenkel and his group, and is mainly transcribed from [Wehenkel and Pavella, 1996, Pavella and Wehenkel, 1998]. It is not a critical survey, and by no means could it do justice to the many contributions in the field which is continuously being fed by a flourishing technical literature. The interested reader may kindly refer to [Wehenkel, 1998].

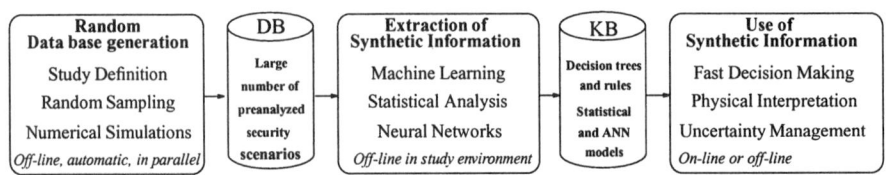

Figure 1.5. AL framework for TSA. Adapted from [Wehenkel and Pavella, 1996]

6.1 Problem statement

Automatic Learning (AL) in general is concerned with the design of automatic procedures able to learn a task on the basis of a learning set of solved cases of this task. Three main families of AL methods may be distinguished, namely : (i) machine learning, a subfield of symbolic artificial intelligence (decision trees are members of this family); (ii) artificial neural network based learning; (iii) statistical pattern recognition and regression.

In the particular context of power system TSA&C, the AL approach may be schematically described by Fig. 1.5 : random sampling techniques are used to screen all relevant situations in a given context (here transient stability), while existing analytical tools are exploited – if necessary in parallel – to derive detailed stability information. The heart of the framework is provided by AL methods used to extract and synthesize relevant information and to reformulate it in a suitable way for decision making. This consists of transforming the data base (DB) of case by case numerical simulations into a power system security *knowledge base* (KB). As suggested in Fig. 1.5, a large variety of AL methods may be used in a toolbox fashion, according to the type of information they may exploit and/or produce. The final step consists of using the extracted synthetic information (decision trees, rules, neural network or statistical approximators) either in real-time, for fast decision making, or in the off-line study environment, so as to gain new physical insight and to derive better system and/or operation planning strategies.

This section briefly reviews various classes of AL methods applied to TSA&C.

6.2 Overview of AL methods

Two broad types of AL may be distinguished : supervised and unsupervised. Supervised learning usually aims at constructing a model for an assumed relationship between input attributes (e.g., real-time measurements) and outputs (e.g., stability margins), while unsupervised learning (or clustering) essentially aims at either uncovering similarities among groups of security scenarios or correlations among groups of variables used to describe such scenarios. In what follows we restrict ourselves to supervised learning. Within this context,

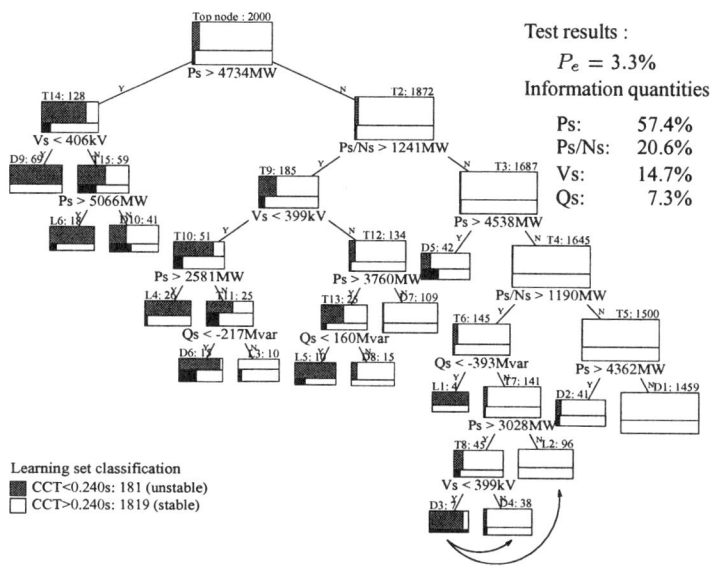

Figure 1.6. DT for transient stability assessment. Taken from [Wehenkel and Pavella, 1996].

three classes of methods are distinguished. They provide three *complementary* types of information as discussed below.

6.2.1 Decision trees (DTs)

We illustrate this method on the DT portrayed in Fig. 1.6 built for the purpose of transient stability [Wehenkel and Pavella, 1996]. The tree comprises "test nodes" and "terminal nodes"; it involves dichotomic tests automatically carried out by the building procedure outlined below.

The tree building starts at the tree's top node, with the entire learning set, which comprises a mixture of learning cases of the various classes. (In Fig. 1.6, 1819 stable and 181 unstable cases).[9] The building procedure consists of identifying that candidate attribute,[10] along with that threshold value of it which allows a decomposition of the mixture of stable-unstable cases into the two most purified learning subsets; these latter are then directed towards the two successors of the top node. This dichotomic procedure is repeatedly applied to each of the successive test nodes, until getting a purified enough subset whose further splitting is deemed statistically meaningless. This subset composes a

---

[9] In addition to these 2000 learning cases, another 1000 cases were used to assess the accuracy of the DT.
[10] "Candidate attributes" are system parameters a priori deemed likely to drive the phenomena of concern. They are suggested by past experience and chosen by the human expert (e.g., here, voltages, generated powers, power flows on important lines). Note that, for reasons appearing below, they are often chosen among pre-disturbance system parameters.

26 TRANSIENT STABILITY OF POWER SYSTEMS

terminal node, labeled stable or unstable according to its majority population; for example, in Fig. 1.6, the leftmost terminal node D9 is an unstable node.

The candidate attributes thus selected at the various test nodes become the "test attributes". Note that a same test attribute may appear at many test nodes, although with different threshold values. For example, in Fig. 1.6 the test attribute P_s (denoting a generated power) appears at the top node as well as at other, lower-level test nodes, with different threshold values (4734, 4538, 5066, ... MW).

Besides identifying the test attributes, the tree also appraises their respective influence on the phenomena, in terms of their "information quantity" (IQ). Note that this is a "measure of purification ability" which takes into account the size of the learning (sub)set of concern. Hence, the higher the position of a test attribute in the tree hierarchy, the larger its IQ − and hence its influence on the phenomena. Finally, note that the IQ of a test attribute which appears more than once on the tree is the sum of its partial IQ's.

The essential outcomes of the above procedure are as follows.

- Selection of the test attributes: these are a (generally small) subset of the candidate attributes proposed to the tree building. They represent the salient parameters driving the phenomena (here, transient stability). In Fig. 1.6, only 4 out of the 17 candidate attributes have thus been selected; they appear at the test nodes of the tree of Fig. 1.6 and are explicitly listed, along with their total IQs in the upper rightmost part of the figure.[11]
- Synthetic description of the physical phenomena of concern. The description follows the very hierarchical structure of the tree: the more essential the part of a test on the mechanism, the higher its position in the tree.
- Straightforward classification of (so far) unseen cases. This is achieved by placing the case at the top tree node and letting it progress down the tree by subjecting it to the successive tests that it thus meets. Eventually, the case reaches a terminal node and is given its class.

To fix ideas, according to Fig. 1.6, a case whose P_s is larger than 4734 MW and whose V_s is smaller than 406 kV will be directed to node D9 and classified as unstable by the tree.

The former two outcomes refer to synthetic information, which yields the INTERPRETABILITY of the phenomena; the latter concerns detailed information, and yields the classification.

Actually, the tree's unique feature and particular strength lies in this interpretability property, which is essential − at least for stability concerns; for that reason, it will also be the backbone of the hybrid(ised) AL approaches advocated below.

---

[11] In Fig. 1.6 these IQs are given in percentages.

As far as classification is concerned, the DT classifies a case by merely relying on the values that the tree test attributes assume for that case. Hence, by choosing as candidate attributes pre-disturbance (and hence predetermined) system parameters, the classification computing effort is reduced merely to the time required by the case to traverse the tree, i.e. to an extremely short time (see Table 1.2). But, on the other hand, it should also be observed that DTs provide a discrete (staircase) rather than smooth approximation. This prevents one from getting back margin values (e.g. CCTs) which could have been used in the data base. In some sense, this discrete type of approximation is the price to be paid in order to get good interpretability.

Another piece of information of great practical importance is the possibility to infer means of control. For example, Fig. 1.6 suggests how to enhance the stability of a case directed to the lowest unstable terminal node (D3): move it either to the next, stable terminal node (D4) by acting on the value of attribute V_s or to the immediately higher-level stable node (L2) by acting on the value of attribute V_s.

To summarize the above discussion, in addition to interpretability, DTs provide means for analysis (classification), sensitivity analysis (influence of the driving parameters on the phenomena) and control, and that in extremely short times (fractions of ms, see in Table 1.2). In short, DTs provide all the necessary ingredients to meet the needs identified in Section 2, and, in particular, the needs of real-time operation.

There are two generalizations of decision trees of interest in the context of TSA: *regression* trees which infer information about a numerical output variable; *fuzzy* trees which use fuzzy logic instead of standard logic to represent output information in a smooth fashion. Both regression and fuzzy approaches allow inferring information about security margins. Fuzzy trees have not yet reached the maturity of crisp classification or regression trees, but they seem particularly well suited to TSA problems. Indeed, they appear to be more robust than classical decision trees and are able to combine smooth input/output approximation capabilities of neural networks with interpretability features of crisp DTs.

6.2.2 Artificial neural networks

In contrast to crisp DTs, ANNs are able to provide continuous margins (e.g. to recover the CCT information contained in the data base), at least in the neighborhood of stability boundaries. This ability is owing to the nonlinear relationships between the ANN inputs and outputs, through the neurons of their hidden layers. But, on the other hand, these complex relationships make it hard to get an insight into the phenomena. Thus, for example, it is (almost) impossible to identify those input attributes which drive the phenomena, and to appraise their influence. In some sense, in contrast to DTs, ANNs provide smooth approximations at the sacrifice of interpretability.

For the stability problems considered here, this "blackbox" type of information is a handicap; it makes ANNs uninteresting to use, unless combined with other, more transparent approaches, like DTs (see below the DT-ANN model).

6.2.3 Statistical pattern recognition

The previous two approaches essentially compress detailed information about individual simulation results into general, more or less global stability characterizations.

Additional information may however be provided in a case by case fashion, by matching an unseen (e.g., real-time) situation with similar situations found in the data base. This may be achieved by defining generalized distances so as to evaluate similarities among power system situations, together with appropriate fast data base search algorithms.

A well known such technique is the "k Nearest Neighbors" (kNN) method able to complete DTs and ANNs. It consists of classifying a state into the majority class among its k nearest neighbors in the learning set. The main characteristics of this method are high simplicity but sensitivity to the type of distances used. In particular, to be practical, ad hoc algorithms must be developed to choose the distances on the basis of the learning set. While in the past this method was generally exploiting a small number of sophisticated ad hoc input features manually selected on the basis of engineering judgment (e.g., in the context of TSA, see the survey by [Prabhakara and Heydt, 1987]), nowadays the emphasis is more on the research of automatic distance design methods exploiting the learning states.

6.2.4 Hybrid AL approaches

Each of the above "pure" AL approaches has its own assets and relative weaknesses; hence the idea of combining them so as to gather their assets while evading their weaknesses. Of course, the choice among various combinations depends on the particular application sought.

Below, two such combinations developed for the purpose of transient stability are outlined. They both use DTs for their interpretability capabilities and for providing test attributes and their information quantity.

DT-ANNs. The DT provides a synthetic view of the phenomena and identifies the relevant parameters (its test attributes); these are then used as input variables to the ANN. The ANN may output either a continuous type of margins or a discrete classification, more accurate than that of the DT.

Table 1.2. Typical performances of AL methods applied to TSA. Adapted from [Wehenkel and Pavella, 1996]

Method	Accuracy %		CPU time (SUN Sparc 20/50)	
	Classif.	Regr.	Off-line (s)	On-line (ms)
Crisp DT	3.3	–	60	0.1
Fuzzy DT	–	1.3	–	1.6
Pure ANNs	–	1.6	33,000	3.6
Hybrid ANNs	2.7	1.4	3,200	1.6
Pure kNN	6.6	6.7	200	100
Hybrid kNN	1.4	1.7	50,000	100

DT-kNNs. The DT provides a synthetic view of the phenomena, including the test attributes, along with their IQ. The kNN uses these attributes to define the attribute space, and their respective IQs to weight them in the distance measure. The DT-kNN model may be further optimized using genetic algorithms, as proposed in [Houben et al., 1997].

Two main outcomes of the DT-kNN combination are: description of the global mechanism of the phenomena, provided by the DT; description of the detailed mechanism concerning an (unseen) case by comparison with its (known) nearest neighbours, provided by the kNN. This latter detailed information may be used in various ways. For example, to suggest means of control, to avoid dangerous errors,[12] and to identify outliers, as discussed earlier. This latter aspect is quite a unique feature of kNNs.

6.3 Performances and assets

6.3.1 Overall comparison

Tables 1.2 and 1.3 give a synoptic view of the performance and salient features of the "pure" and "hybrid" AL approaches considered above. The accuracy is expressed in terms of test set error rate, P_e (in %). Table 1.3 gathers general comparative properties. The assessments rely on the transient stability problem and the data base used to build the DT of Fig. 1.6. Recall that actually 2000 cases of this data base were used to learn the various AL methods, the other 1000 cases were used to assess their accuracy, or their corresponding test set error rate, P_e.

---

[12] i.e. dangerous misclassifications; such a dangerous error arises when an actually (very) unstable case is declared stable by the classifier.

Table 1.3. Salient features of AL methods applied to TSA. Adapted from [Wehenkel and Pavella, 1996]

	Method	Functionalities	Computational: Off-line	On-line
Pure	Crisp DTs	**Good interpretability (global)**. Discrete. Good accuracy for simple "localized" problems. Low accuracy for complex, diffuse problems.	Very fast	Ultra fast
	ANNs	**Good accuracy**. Low interpretability. Possibility for margins and sensitivities.	Very slow	Fast
	kNN	**Good interpretability (local)**. Conceptual simplicity.	Very slow	Very slow
Hybrid	Fuzzy DTs	**Good interpretability (global)**. Symbolic and continuous. More accurate and more robust than crisp trees. Possibility for margins and sensitivities.	Slow	Very Fast
	DT-ANN	Combine features of DTs and ANNs	Slow	Very Fast
	DT-kNN	Combine features of DTs and kNNs	Slow	Slow

Observe that the best accuracy performance for classification is provided here by the hybrid kNN (last row of Table 1.2). Actually, this is a kNN-DT-GA model where the genetic algorithm (GA) allows the accuracy of kNN-DT to be enhanced at the expense of lengthy computations.

6.3.2 Main assets of AL methods

The foregoing discussions identify, and Table 1.3 highlights, three salient assets intrinsic to (some of the) AL methods over the deterministic ones, namely:

- interpretability of the phenomena;
- extraordinary on-line computational efficiency;
- management of uncertainties.

The former two properties are linked to the models used by the AL methods. Management of uncertainties is obtained thanks to the way of generating a data base, where the uncertain information is (suitably) randomized. This latter property is of great importance in real-world problems where uncertainties are always present. They are inherent to the necessarily limited knowledge of the power system modeling (e.g., modeling of loads connected to high voltage buses) and parameters, not to mention the lack of information due to economic-political factors.

6.4 Comparison of methods

Paragraph 2.3 has identified the present needs of transient stability assessment, in the context of preventive and emergency modes.

Sections 4 to 6 have described the three classes of existing approaches to transient stability assessment, highlighted their potentials and identified the needs that each one of them is able to encounter. It was suggested that some of the needs may be met by more than one method in a complementary rather than competitive way.

A comparison will be proposed at the end of the monograph, where the various techniques will be assessed in terms of various evaluation criteria, such as: modeling possibilities; type of information required; off-line preparation tasks; real-time computational requirements; type of information provided.

7. SCOPE OF THE BOOK

This monograph is devoted to a comprehensive and unified approach to TSA&C: the hybrid temporal-direct method called SIME. The approach is comprehensive, since it covers both preventive aspects (be it for planning, operation planning or real-time operation) and emergency aspects. The approach is unified, since it relies on the same basic method suitably adapted to the various requirements.

The scope of the book is essentially threefold. First, as a textbook, to give a detailed description of SIME so as to make it accessible to practicing engineers, researchers, final year undergraduate and PhD students. Second, to illustrate its outcomes by concrete real-world examples in the various application contexts, so as to guide the choice among existing strategies for transient stability assessment and control; further, to open avenues for new operational stragegies. Third, to provide researchers and PhD students with real-world topics for further investigation.

The monograph benefits from authors' various collaborations with electric industry, research institutes, and EMS (energy management system) constructors. These collaborations have guided the development of appropriate software functions. At the same time, they gave the opportunity to test the developed functions on a large variety of real power systems; they thus contributed to make the method robust, reliable, and fully operational.

N.B. The illustrations on the simple three-machine system may easily be checked by the reader, using for example MATLAB together with the data provided in Appendix B. Illustrations on real-world power systems aim at exploring/describing transient stability phenomena specific to large power systems.

8. SUMMARY

The transient stability problem has first been introduced in the realm of power system security, and its specifics along with its study contexts have been identified. Further, the need for tailor-made techniques in the various study contexts, and especially in real-time operation, have been specified.

The impact of the restructuring electric industry has next been considered. It was pointed out that the new market environment creates needs considerably more stringent than those of the vertically organized electric sector.

The possibilities of the conventional time-domain approach have then been scrutinized in the light of the resulting requirements; its fundamental role has thus been highlighted but also its serious deficiencies.

Two major classes of non-conventional approaches have next been reviewed: the direct or hybrid direct–temporal methods, which are deterministic in essence, and the probabilistic, automatic learning methods. The review helped identify their complementary features and strengths.

In short, this chapter has provided material necessary to make the monograph self-reliant.

Chapter 2

INTRODUCTION TO SIME

The objectives of this chapter are:

- *to trace the origins of SIME and state its principle (Section 1)*
- *to define the rules of OMIB's automatic identification and set its general formulation (Section 2)*
- *to scrutinize the derivation of stability margins and related issues (Section 3)*
- *to illustrate the developments on a simple example easily tractable by the reader (Section 4), with the exception of phenomena specific to large-scale systems (Section 5)*
- *to give a flavor of the way the techniques developed in this chapter will be used in the remainder of the monograph (Section 6)*
- *to point out main differences between preventive and emergency SIME (Section 7).*

1. FOUNDATIONS
1.1 OMIB: concept and variants

As mentioned in Chapter 1, SIME belongs to the general class of transient stability methods which rely on a one-machine infinite bus (OMIB) equivalent. To get a better insight into its specifics, it is therefore interesting to take a look at the general class of OMIB methods.

All OMIB-based methods rely on the observation that the loss of synchronism of a multimachine power system originates from the irrevocable separation of its machines into two groups, which are replaced by a two-machine system and then by an OMIB equivalent. Thus, an OMIB may be viewed as a transformation of the multidimensional multimachine dynamic equations into a single dynamic equation. This latter takes on various forms, depending upon the power system modeling and the assumed behaviour of the machines within each group. We distinguish three types of OMIBs: "time-invariant", "time-varying" and "generalized" ones.

The foundations of all these OMIBs rely on the classical, well-known equal-area criterion (EAC). EAC along with related notation is recalled in Appendix A.

1.1.1 Time-invariant OMIB

A time-invariant OMIB is obtained under the following assumptions: (i) simplified power system modeling; (ii) coherency of the machines within each one of the two groups, so as to "freeze" their relative motion in the fault-on and the post-fault periods. The dynamic equations of a multimachine system may therefore be transformed into an equivalent OMIB equation of the form

$$M\ddot{\delta} = P_a = P_m - P_e = P - P_{\max}\sin(\delta - \nu) \quad (2.1)$$

where M, P, $P_{\max}$ and ν take on constant values (different in the fault-on and the post-fault configurations); hence the name "time-invariant" OMIB. Note that eq. (2.1) is similar to that of the simple one-machine infinite bus system (see Appendix A). It describes the typical sinusoidal variation on the $P_a - \delta$ plane, to which the well-known EAC may apply.

The first publications in the area were the "worst case" approach by Rahimi and Schaffer [Rahimi and Schaffer, 1987], and the EAC (Extended Equal-Area Criterion) by Xue et al. [Xue et al., 1986, Xue, 1988, Xue et al., 1988].

1.1.2 Time-varying and generalized OMIBs

If we relax the coherency assumption while keeping the simplified power system model, we get "time-varying" OMIBs: P, $P_{\max}$, ν are no longer constant, and the dynamics in eq. (2.1) become piece-wise sinusoidal. Such time-varying OMIBs were for instance used in the so-called GEAC (for Generalized EAC) [Rahimi, 1990] and DEEAC (for Dynamic EEAC) [Xue and Pavella, 1993, Xue et al., 1993, Pavella and Murthy, 1993].

If, in addition, we consider detailed power system models, we come up with what we will call the "generalized" OMIB. Its dynamic model is still expressed by

$$M\ddot{\delta} = P_m - P_e = P_a \ . \quad (2.2)$$

But, here, the $P_a - \delta$ variation is no longer sinusoidal; nevertheless, the energy concept of EAC still holds valid.

Various approaches have been proposed in recent years to build up generalized OMIBs; for example: "mixed" DEEAC [Zhang, 1993], HEEAC (for hybrid EEAC) [Zhang et al., 1995], IEEAC (for Integrated EEAC) [Xue, 1996], FASTEST [Xue et al., 1997], SIME [Zhang et al., 1997a, Zhang et al., 1998]. The various approaches differ in many respects (e.g. in the way of identifying the mode of machines' separation, of updating the OMIB parameters, of computing the stable and unstable margins, of assessing stability limits); but they all rely on the same concept viz., the generalized OMIB transformation.

1.2 From EEAC to SIME

SIME (for SIngle Machine Equivalent) is a transient stability method based on a generalized OMIB. More precisely, it is a hybrid, temporal-direct method: temporal, since it relies on the multimachine system evolution with time; direct, like the EEAC, from which it originates.

To justify, or at least explain the reasoning leading to EEAC, let us go back to the discussions of §§ 5.2.3, 5.3.2 of Chapter 1. It was said that in the context of $V-$ functions constructed for the transient stability problem, the difficulties linked to the stability domain estimation vanish when considering two-machine or one-machine systems. Based on this observation, two ways were suggested to overcome the stability domain estimation problems in the case of multimachine power systems. The first was the Bellman decomposition-aggregation method [Bellman, 1962] for dynamic systems which if applied to power systems translates as: decompose the system into two-machine (or one-machine) subsystems; construct scalar $V-$ functions for these latter; consider these scalar functions as the components of a vector Lyapunov function; finally, aggregate the subsystems to infer stability information about the original, multimachine system [Grujic et al., 1987]. But, as mentioned in Chapter 1, this approach fails in practice, essentially because of the overly stringent stability conditions imposed on the aggregated system. The second way to circumvent the difficulty was the construction of an OMIB system itself. As mentioned earlier, this is on the basis of EEAC.

EEAC (for Extended Equal-Area Criterion) is a direct transient stability method relying on a time-invariant OMIB. The crux of this approach, as in any OMIB-based approach, is the identification of the right decomposition pattern of the system machines into two groups.

The first attempts to reduce the multimachine trajectory to that of an OMIB date back to the eighties. They originated from two distinct research groups which proposed almost simultaneously the OMIB concept that, however, developed in different ways [Rahimi and Schaffer, 1987, Xue et al., 1986, Xue, 1988, Xue et al., 1988]. (A short account may be found in [Pavella and Murthy, 1993].)

The fundamental difference between the original version of EEAC and SIME is that EEAC relies on a time-invariant OMIB that constructs by assuming the classical simplified machine and network modeling and by "freezing" once and for all the machine rotor angles at t_0, the initial time of the disturbance inception. As a consequence, EEAC is a pure direct method, free from any transient stability program; it thus yields analytical expressions that are extraordinarily fast to compute, but introduces approximations about the machines coherency and their (over)simplified modeling.

The EEAC started being tested on the French EHV power system in the early nineties. The first results were very encouraging [Xue et al., 1992]. At the same time, they clearly revealed that the efficient identification of the correct decomposition pattern of the machines was a difficult but crucial task for the validity of the method and its practical application to real systems. The difficulty increases when dealing with highly meshed systems. Solving this problematic but key issue [Xue and Pavella, 1993] has contributed to make EEAC robust and able to provide an interesting alternative to the conventional Time-Domain (T-D) method, by alleviating the bulky simulations involved in planning studies [Dercle, 1995].

The good performances of the method reinforced engineers' interest in its use as a tool for operational planning and real-time operation. At the same time, engineers started being more demanding as concerning applicability limitations of the method; in particular, they started considering its modeling limitations as a serious handicap. Relaxing EEAC from these and some other related constraints became a strong motivation for further research. This has gradually led to the development of "dynamic" and of "hybridized" EEAC versions [Xue et al., 1993, Zhang et al., 1995]; little by little, these successive versions departed from the original EEAC idea and resulted in the SIME method [Zhang et al., 1996, Zhang et al., 1997a, Bettiol et al., 1997, Pavella et al., 1997, Zhang et al., 1998].

The first version of SIME was using temporal information of the multimachine power system, furnished step-by-step by a T-D program.

This SIME coupled with T-D programs has subsequently been (re)named Preventive SIME, to distinguish it from the Emergency SIME, proposed more recently, which uses real-time measurements collected on the system power plants. As suggested by these names, the Preventive SIME operates in the preventive mode, prior to any disturbance inception, while the Emergency SIME aims at controlling the power system after a disturbance inception, so as to prevent loss of synchronism. More about Preventive vs Emergency SIME will be found in Section 7 of this chapter, after a description of SIME's essentials.

Indeed, the scope of this chapter is to state SIME's general principle, set up its basic formulation and related material, illustrate main features, and finally describe the overall organization of the book, thus guiding the reader through the various subject matters which cover the whole field of power system transient stability assessment and control.

1.3 Principle

Like all OMIB-based methods, SIME relies on the following two propositions.

Proposition 1. However complex, the mechanism of loss of synchronism in a power system originates from the irrevocable separation of its machines into two groups: one composed of the "critical machines" (CMs), which are responsible of the loss of synchronism, the other of the "non-critical machines" (NMs). Hence, the transient stability behaviour of the multimachine system may be inferred from that of an OMIB properly derived from the above decomposition pattern into two groups.[1]

Proposition 2. The stability properties of an OMIB may be inferred from EAC constructed for this OMIB.

More specifically, SIME uses a generalized OMIB whose parameters are inferred from the multimachine temporal data and refreshed at the same rate. These data are obtained either from time-domain transient stability simulations of anticipated contingencies (Preventive SIME), or from real-time measurements reflecting the actual transient stability behaviour of a power system, subjected to a contingency (Emergency SIME). In either case, SIME may be viewed as a means of compressing multimachine data to extract information about *transient stability margins and critical machines*. As will be shown below, these two pieces of information broaden dramatically the possibilities of SIME with respect to the temporal multimachine information. In particular, they open avenues to sensitivity analysis and control. At the same time, by updating this compressed information at the rate of acquisition of the multimachine parameters, SIME preserves the accuracy of the multimachine information.

2. GENERAL FORMULATION

Basically, SIME concentrates on the post-fault configuration of a system subjected to a disturbance which presumably drives it to instability, in order to: identify the mode of separation of the machines[2] into two groups; replace these latter successively by two machines, then by OMIB; assess the transient

---

[1] Note that the machines of each group are by no means assumed to behave coherently. They even may split subsequently into sub-groups which would themselves cause loss of synchronism if they were considered separately from the other group. The point is that the system loses synchronism as soon as the first major machines separation occurs.

[2] The terms "mode of machines' separation" and "mode of instability" will be used interchangeably. They define the "critical OMIB", i.e., the OMIB of concern, or, simply, the OMIB.

stability properties of this OMIB via the energy concept of EAC [Zhang et al., 1997a, Zhang et al., 1998].

The various steps of the method are elaborated below and illustrated in Figs 2.1 to 2.3 corresponding to the small 3-machine system described in Section 1 of Appendix B.

Without loss of generality, the formulation and illustrations of this chapter are developed in the context of the Preventive SIME.

2.1 Critical machines identification

The notion of critical machines is intimately related to unstable scenarios. By definition, on such an unstable multimachine trajectory the critical machines (CMs) are those which go out of step i.e., which cause the system loss of synchronism. To identify them, SIME drives the T-D program first in the during-fault then in the post-fault configuration; and as soon as the system enters the post-fault phase, SIME starts considering, at each time step of the program, candidate decomposition patterns, until one of them reaches the instability conditions (2.15) defined by EAC (see § 2.4).

More precisely, at each time step of the post-fault simulation, SIME sorts the machines according to their rotor angles, identifies the very first larger rotor angular deviations ("distances") between adjacent machines, and considers as *candidate* CMs those which are above each one of these larger distances. The corresponding candidate OMIB parameters are computed according to expressions (2.3) to (2.12) of § 2.2. The procedure is carried out until a candidate OMIB reaches the unstable conditions defined by (2.15): it is then declared to be the critical OMIB, or simply the OMIB (of concern).

ILLUSTRATION. Let us illustrate the OMIB identification procedure on the 3-machine system, supposed to be subjected to contingency Nr 2.[3] The simulations are performed by SIME-MATLAB (see Appendix B).

Figure 2.1a portrays the machine swing curves from $t = t_0 = 0$ (time of fault inception) to $t_e = 117$ ms (time of fault clearance) and from t_e to $t_u = 458$ ms ("time to instability"). Time t_u is determined according to the developments of § 2.4 and corresponds to the time when the critical OMIB loses synchronism.

At $t = t_u$, one reads: $\delta_{m_2} = 127.5°$; $\delta_{m_3} = 81.6°$; $\delta_{m_1} = -45°$. Accordingly, the angular distances are: $\delta_{m_2} - \delta_{m_3} = 45.9°$; $\delta_{m_3} - \delta_{m_1} = 126.6°$. They yield two candidate OMIBs: the one is composed of 1 CM and 2 NMs (δ_{m_2} and δ_{m_3}, δ_{m_1} respectively); the other is composed of 2 CMs and 1 NM (δ_{m_2}, δ_{m_3} and δ_{m_1} respectively). Anticipating, we mention that, according to § 2.4, the actual critical OMIB is composed of 2 CMs and 1 NM. Note that, here, this OMIB corresponds to the largest "distance" (angular deviation between adjacent machines at t_u); but this is not a general rule. Figure 2.1a plots also the OMIB swing curve computed from the

---

[3]The 3-machine system as well as the considered contingencies are fully described in Section 1 of Appendix B.

3-machine swing curves, according to eqs (2.3) to (2.6) of § 2.2. Note that, here, the OMIB trajectory is more advanced than that of the most advanced machine: $\delta > \delta_{m_2}$; in particular, $\delta(t_u) = 158.0°$ (vs $\delta_{m_2}(t_u) \equiv 127.5°$).

Further, using the formulas developed in the following paragraph, Fig. 2.1b sketches the OMIB $P - \delta$ plots, computed according to eqs (2.3) to (2.12). With reference to the shape of P_e and P_m curves of Fig. 2.1b, note that because of the machines modeling, the P_{eP} curve is not sinusoidal as in the pure EAC representation; but still looks alike. On the other hand, the plot of P_m curve still remains a straight line, suggesting that there is no fast valving nor fast acting governors modeled here.

Finally, Fig. 2.1c provides a general notation, used throughout the monograph.

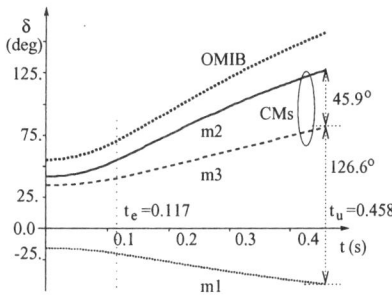

(a) Critical machines and OMIB trajectories
Nr of CMs: 2
Nr of NMs: 1

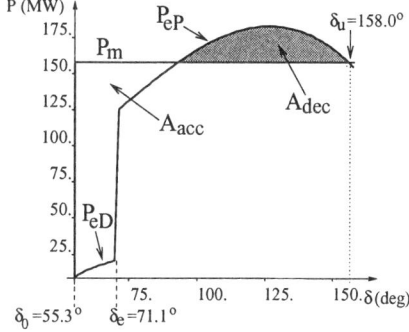

(b) Power-angle OMIB representation.
Resulting EAC parameters:
$\delta_e = 71.1°$; $\delta_u = 158.0°$ → $t_u = 0.458$ s.

(c) Notation:

P_e : electrical power ; P_m : mechanical power ;
$P_a = P_m - P_e$: accelerating power
t_e : clearing time ; t_u : time to instability
$\delta_e = \delta(t_e)$: clearing angle ; $\delta_u = \delta(t_u)$: unstable angle
Subscript D stands for "during-fault" (or fault-on) configuration
Subscript P stands for "post-fault" configuration

Figure 2.1. Swing curves and OMIB $P - \delta$ representation of the 3-machine system. Contingency Nr 2, $t_e = 117$ ms. General notation

2.2 Derivation of OMIB time-varying parameters

As already mentioned, the OMIB transformation results from the decomposition of the system machines into two groups, the aggregation of these latter into their corresponding center of angle (COA) and finally the replacement of the two COAs by an OMIB. These transformations use the parameter values of the system machines and are refreshed at the rate these values are furnished by the step-by-step T-D program. The resulting time-varying OMIB parameters are thus a faithful picture of the multimachine parameters.

The formulas developed below correspond to the pattern which, presumably, decomposes the machines into critical machines (CMs, subscript C) and non-critical machines (NMs, subscript N). However, they may also apply to any other (candidate) decomposition pattern.

The expressions of corresponding OMIB parameters δ, ω, M, P_m, P_e, P_a are derived as follows.

(i) Denoting by $\delta_C(t)$ the COA of the group of CMs, one writes:

$$\delta_C(t) \triangleq M_C^{-1} \sum_{k \in C} M_k \delta_k(t) \ . \tag{2.3}$$

Similarly:

$$\delta_N(t) = M_N^{-1} \sum_{j \in N} M_j \delta_j(t) \ . \tag{2.4}$$

In the above formulas:

$$M_C = \sum_{k \in C} M_k \ ; \ M_N = \sum_{j \in N} M_j \ . \tag{2.5}$$

(ii) Define the rotor angle of the corresponding OMIB by the transformation

$$\delta(t) \triangleq \delta_C(t) - \delta_N(t) \ . \tag{2.6}$$

The corresponding OMIB rotor speed is expressed by

$$\omega(t) = \omega_C(t) - \omega_N(t) \tag{2.7}$$

where

$$\omega_C(t) = M_C^{-1} \sum_{k \in C} M_k \omega_k(t) \ ; \ \omega_N(t) = M_N^{-1} \sum_{j \in N} M_j \omega_j(t) \ . \tag{2.8}$$

(iii) Define the equivalent OMIB mechanical power by

$$P_m(t) = M \left(M_C^{-1} \sum_{k \in C} P_{mk}(t) - M_N^{-1} \sum_{j \in N} P_{mj}(t) \right) \ , \tag{2.9}$$

the equivalent OMIB electric power by

$$P_e(t) = M \left(M_C^{-1} \sum_{k \in C} P_{ek}(t) - M_N^{-1} \sum_{j \in N} P_{ej}(t) \right) \ , \tag{2.10}$$

and the resulting OMIB accelerating power by

$$P_a(t) = P_m(t) - P_e(t) . \tag{2.11}$$

In the above expressions, M denotes the equivalent OMIB inertia coefficient

$$M = \frac{M_C M_N}{M_C + M_N} . \tag{2.12}$$

2.3 Equal-area criterion revisited

As recalled in Appendix A, the well-known EAC proposed round the thirties [Dahl, 1938, Skilling and Yamakawa, 1940, Kimbark, 1948] was dealing with a single machine connected to an infinite bus and modeled in the standard oversimplified way. Its purpose was to determine the stability properties of the single-machine infinite bus system without solving the swing equation formally expressed by (2.1). In other words, EAC was able to determine the maximum excursion of δ and hence the stability of the system without computing the time response through formal solution of the swing equation.

Despite the above extremely restrictive applicability conditions, EAC has been one of the most powerful tools in transient stability studies, thanks to the extraordinary insight it provides into the stability phenomena and their interpretations. Certainly, detailed transient stability analysis relies, one way or the other, on T-D methods where the non-linear sets of differential and algebraic equations (1.1), (1.2) are solved step-by-step by using numerical integration techniques. But despite the extraordinary performances of T-D methods achieved thanks to the tremendous progress in computers, EAC remains a unique tool for sensitivity analysis and control issues. This is why it always retains due attention in textbooks and publications dealing with transient stability.

EAC relies on the concept of energy.[4] In short, it states that the stability properties of a contingency scenario may be assessed in terms of the stability margin, defined as the excess of the decelerating over the accelerating area of the OMIB $P - \delta$ plane:

$$\eta = A_{dec} - A_{acc} . \tag{2.13}$$

In the above expression (2.13), the accelerating area represents the kinetic energy stored essentially during the fault-on period, while the decelerating area

---

[4] Although not surprising, it is interesting to note that the EAC coincides with the Lyapunov criterion using the energy-type Lyapunov function mentioned in Chapter 1, when constructed for the same single-machine infinite bus system and the same classical model (e.g., see [Pavella and Murthy, 1993]).

represents the maximum potential energy that the power system can dissipate in the post-fault configuration. Accordingly, the OMIB system is transient stable if the accelerating area is smaller than the maximum decelerating area; stated otherwise, the OMIB system will be unstable if $\eta < 0$, stable if $\eta > 0$, borderline (un)stable if $\eta = 0$.

Thus, from a physical point of view, eq. (2.13) states that [Zhang, 1995]: if the stored kinetic energy can be released as potential energy in the post-fault system configuration, the system will be stable; otherwise it will be unstable.

Note that since this statement expresses the conservation of energy, it is fully general and may apply to any OMIB system.[5] In particular, it applies to the generalized OMIB whose dynamics is expressed by (2.2). The main and essential difference with the "original" EAC is that SIME computes the $P_m - \delta$ and $P_e - \delta$ curves of the generalized OMIB from the multimachine data provided by a T-D program by solving eqs (1.1), (1.2); the computation is needed only for the (generally short) duration determined by EAC. Depending upon the stability case, this duration reduces to the time to reach the OMIB angle δ_u or δ_r, as appropriate, defined as follows.[6]

- the unstable angle, δ_u, (u for "unstable") is found at the crossing of the P_{eP} and P_m curves for an unstable scenario
- the stable (or return) angle, δ_r, (r for "return") represents the maximum angular excursion of P_{eP} for a stable scenario[7]
- $\delta_u = \delta_r$ when a maximum angular excursion is reached at the crossing of P_{eP} and P_m.

Figures 2.1b and 2.2b are relative respectively to an unstable and a stable scenario under same operating conditions and same contingency, apart from its clearing time. Using the notation of these figures, we can express the stability margin (2.13) as

$$\eta = -\int_{\delta_0}^{\delta_{ch}} P_a d\delta - \int_{\delta_{ch}}^{\delta_u} P_a d\delta = -\int_{\delta_0}^{\delta_u} P_a d\delta . \qquad (2.14)$$

Angle δ_{ch} ($\delta_u > \delta_{ch} > \delta_0$) denotes the angle where P_a changes sign (from positive to negative). Note that this angle does not necessarily coincide with δ_e, i.e., with the switch from P_{eD} to P_{eP} curves as is normally the case of the original EAC (see Appendix A); hence, the accelerating power P_a may change sign at $\delta_{ch} \neq \delta_e$.

---

[5] Of course, to make sense, this OMIB representation has to be valid from a physical point of view.
[6] The analytical definitions of δ_u and δ_r are given below in § 2.4.
[7] This applies to first-swing stability only. Multiswing phenomena will be discussed in Section 5.

For example, in Fig. 2.1b, $\delta_{ch} = 86.0°$: P_a is thus positive from 55.3° to 86.0°, and negative from 86.0° to $\delta_u = 158.0°$.

Formula (2.14) expresses the stability margin in terms of the OMIB $P_a - \delta$ variation in both during- and post-fault configurations. More convenient expressions will be elaborated in Section 3.

But before, let us state the EAC stability conditions in terms of the OMIB time-varying parameters. The developments are illustrated on the previous concrete example, described in Figs 2.1 as well as in Figs 2.2 and 2.3.

ILLUSTRATION. Figures 2.2 report on simulation results obtained under same operating conditions and same contingency with Figs 2.1. The only difference is that, here, the contingency clearing time is small enough ($t_{e2} = 92$ ms vs $t_{e1} = 117$ ms in Figs 2.1) to avoid system's loss of synchronism, yet close enough to t_{e1} to allow considering by continuation the same separation pattern and resulting OMIB. Note that for this stable case the T-D simulations are conducted on the entire simulation period, in order to make sure that the system is not only first-swing stable but also multiswing stable.

Finally, Figs 2.3 gather on their upper parts the power-angle representations in the unstable and stable cases, on their lower parts the corresponding time-evolution of OMIB rotor angle and speed. Note that the upper part of Fig. 2.3b is a truncated version of Fig. 2.2b drawn up to a time slightly longer than the "return time" $t_r = 663$ ms . Notation relative to the small area representing the stable margin in the upper part of Fig. 2.3b is discussed in Section 3.

2.4 Stability conditions

In this paragraph, the stability conditions resulting from the application of eqs (2.13), (2.14) are restated in terms of the OMIB time-varying parameters (angle, speed and accelerating power).

2.4.1 Conditions of unstable OMIB trajectory

An unstable case corresponds to $\eta < 0$, i.e., to $A_{dec} < A_{acc}$ (see eq. (2.13)). Inspection of Fig. 2.1b (or upper part of Fig. 2.3a) suggests that, in such a case, the curve P_{eP} crosses P_m or, equivalently, the accelerating power P_a passes by zero and continues increasing. From a physical point of view, $P_a = 0$ takes place at $\delta = \delta_u$ and marks the OMIB loss of synchronism.

The above reasoning yields the following statement.

An unstable OMIB trajectory reaches the unstable angle δ_u at time t_u as soon as:

$$P_a(t_u) = 0 \ , \ \dot{P}_a(t_u) = \left.\frac{dP_a}{dt}\right|_{t=t_u} > 0 \qquad (2.15)$$

with $\omega > 0$ for $t > t_0$. Whenever met, conditions (2.15) determine the "early termination conditions" of the T-D program. Recall that these conditions are also used to identify the critical OMIB (see § 2.1).

44 TRANSIENT STABILITY OF POWER SYSTEMS

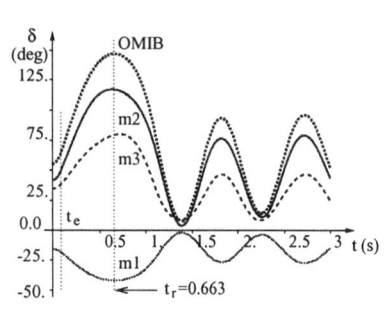
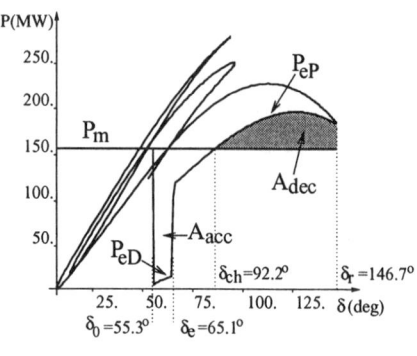

(a)
Stabilized behaviour of the 3-machine system
and corresponding OMIB

(b)
Corresponding power-angle representation of
the OMIB identified under prefault conditions
of Figs 2.1. Resulting EAC parameters:
$\delta_e = 65.1°$; $\delta_r = 146.7°$ ⇒ $t_r = 663$ ms

Figure 2.2. Swing curves and OMIB $P - \delta$ representation of the 3-machine system.
Contingency Nr 2; $t_e = 92$ ms.

2.4.2 Conditions of stable OMIB trajectory

A stable case corresponds to $\eta > 0$, i.e., to $A_{dec} > A_{acc}$. Inspection of Fig. 2.2b and upper part of Fig. 2.3b suggests that in such a case, the acquired kinetic energy is less than the maximum potential energy: P_{eP} stops its excursion at $\delta = \delta_r$, before crossing P_m. Stated otherwise, at $\delta = \delta_r$, $\omega = 0$ with $P_a < 0$, δ stops increasing then decreases.

The above reasoning yields the following statement: a stable OMIB trajectory reaches the return angle δ_r ($\delta_r < \delta_u$) at time t_r as soon as

$$\omega(t_r) = 0 \ , \text{ with } P_a(t_r) < 0. \tag{2.16}$$

Conditions (2.16) are the "early termination stable conditions": they imply that the system is first-swing stable and that the T-D simulations can be stopped, unless multiswing instability phenomena are sought.

2.4.3 Borderline conditions of OMIB trajectory

By definition, a critically stable OMIB trajectory reaches the return angle δ_r with $P_a(t_r) = 0$; it coincides with the critically unstable OMIB trajectory identified by the unstable angle δ_u with $\omega(t_u) = 0$. Stated otherwise, the borderline (un)stable conditions are met when the return angle δ_r equals δ_u. Let δ_c denote this particular, "critical angle" : $\delta_r = \delta_u = \delta_c$.

Observe that in this borderline case, δ_u represents the unstable *equilibrium* angle, since $\omega_u \equiv \omega(t_u) = 0$.

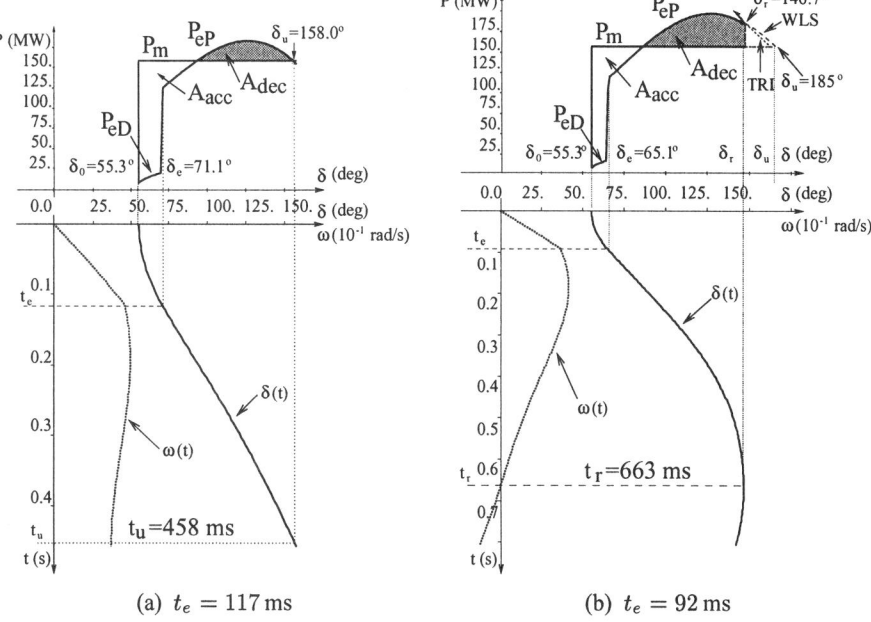

Figure 2.3. OMIB $P - \delta$ and time-domain representations of the 3-machine system. Contingency Nr 2

2.4.4 Objectivity of the stability criteria

Conditions (2.15), (2.16) are fully objective stopping criteria, contrary to the pragmatic criteria used with T-D programs to decide whether a stability simulation has definitely been shown to be (un)stable and can be stopped.[8]

3. STABILITY MARGINS

3.1 Unstable margin

Observing that the OMIB becomes unstable as soon as it meets conditions (2.15), and multiplying both members of the dynamic equation (2.2) by $\dot{\delta}$, we get

$$M \ddot{\delta} \dot{\delta} = P_a \dot{\delta} \,. \tag{2.17}$$

---

[8]This, however, does not imply that the values of the stability limits (critical clearing times or power limits) computed by the two methods will be different.

Further, observing that $\omega_0 = 0$ [9], we get by integration

$$\frac{1}{2}M\omega_u^2 = \int_{\delta_0}^{\delta_u} P_a d\delta. \qquad (2.18)$$

Comparing this equation with eq. (2.14) yields the unstable margin as

$$\eta_u = -\frac{1}{2}M\omega_u^2 \qquad (2.19)$$

where subscript u holds for "unstable" and $\omega_u \equiv \omega(\delta = \delta_u)$.

Remarks
1.- The above expression of the unstable margin is extraordinarily simple and straightforward to compute. It gets rid of the computation of the accelerating and decelerating areas of eq. (2.13) and of the integral in (2.14). It merely requires knowledge of the OMIB inertia coefficient and rotor speed.

2.- Under very unstable simulation conditions, it may happen that an OMIB post-fault trajectory has only positive accelerating power, i.e. no unstable angle δ_u and therefore no margin. Paragraph 3.4 identifies the conditions under which such a situation arises and shows how to circumvent this difficulty by using another type of margin, different from the one defined by EAC.

3.2 Stable margin

Observing that the OMIB remains stable if P_{eP} "returns back" before crossing P_m (see conditions (2.16)), rewriting eq. (2.18) as

$$\eta = -\int_{\delta_0}^{\delta_r} P_a d\delta - \int_{\delta_r}^{\delta_u} P_a d\delta \qquad (2.20)$$

and observing that $\omega = 0$ for $\delta = \delta_0$ and for $\delta = \delta_r$, we get the corresponding stable margin as

$$\eta_{st} = -\int_{\delta_r}^{\delta_u} P_a d\delta = \int_{\delta_r}^{\delta_u} |P_a| d\delta \qquad (2.21)$$

where subscript st stands for "stable".

3.2.1 Remark

Unlike unstable margin which takes on the closed form expression (2.19) and can be computed precisely, stable margin expressed by (2.21) can only be

---

[9] Indeed, $\omega_0 \equiv \omega(t_0 = 0) = 0$, since it represents the pre-fault stable equilibrium solution.

approximated. Indeed, neither δ_u nor $P_{eP}(\delta)$ ($\delta_u > \delta > \delta_r$) can directly be computed, since the OMIB $P_{eP} - \delta$ curve "returns back" at $\delta = \delta_r$ (see Fig. 2.3b).

Two types of approximations are proposed below: one relies on the construction of a triangle and computation of its area, the other on the prediction of the $P_a(\delta)$ curve in the interval $[\delta_r, \delta_u]$ and computation of the integral (2.21).

3.2.2 Triangle approximation

This approximation consists of using a linearized trajectory in the $P_a - \delta$ plane, in the interval $[\delta_r, \delta_u]$. Denoting by P_{ar} the value of P_a at $\delta = \delta_r$, this yields

$$\eta_{st} = \frac{1}{2}|P_{ar}|(\delta_u - \delta_r) \tag{2.22}$$

and corresponds to the small decelerating area portrayed in the upper part of Fig. 2.3b, denoted "TRI".

Note that the validity of the above approximation relies on the following two conditions:

1. the stable angle in eq. (2.22), δ_r, is close enough to δ_u; otherwise, the approximate expression (2.22) may introduce large errors

2. to compute the unstable angle, δ_u, accurately enough, two unstable simulations are needed, in order to use the linear extrapolation formula:

$$\delta_u|_{\eta=0} = \delta_u(k) - \frac{\eta_u(k)}{S_{\delta_u}^{\eta_u}} \tag{2.23}$$

where the sensitivity coefficient $S_{\delta_u}^{\eta_u}$ is expressed by

$$S_{\delta_u}^{\eta_u} \triangleq \frac{\Delta \eta_u}{\Delta \delta_u} = \frac{\eta_u(k) - \eta_u(k-1)}{\delta_u(k) - \delta_u(k-1)} . \tag{2.24}$$

The above expressions comply with several sensitivity definitions (see below § 3.2.5 and Chapter 3).

On the other hand, the reason for searching the angle δ_u corresponding to $\eta = 0$ is clearly suggested by the triangle displayed in the upper part of Fig. 2.3b. Indeed, as can be seen, it would be for that angle δ_u that the curve $P_e(\delta)$ would reach P_m then would immediately start "returning back". Figures 2.5b and 2.5c of § 3.3.2 show that for $\eta = 0$, $\delta_u = \delta_r = \delta_c$.

48   TRANSIENT STABILITY OF POWER SYSTEMS

3.2.3   Weighted least-squares approximation

The weighted least-squares (WLS) approximation was initially developed in the context of Emergency SIME.[10] It consists of extrapolating the $P_a(\delta)$ curve from δ_r to δ_u :

- either by writing

$$P_a(\delta) = a\delta^2 + b\delta + c \qquad (2.25)$$

and solving for a, b, c based on P_a values taken at three successive time steps

- or by using a WLS approximation based on more than three (actually a large number of) time step values.

Angle δ_u is then found by solving eq. (2.25) and taking the solution $\delta > \delta_r$. Finally, η_{st} is computed via eq. (2.21). In the upper part of Fig. 2.3b, the area thus computed is labeled "WLS". More about this predicted $P_a(\delta)$ curve will be found in Chapter 6.

3.2.4   Triangle vs WLS approximation

The triangle approximation relies on two unstable simulations which, in addition, should be quite close to the stable one. When these conditions are met, the resulting computation of the triangle-based positive margin is very accurate.

On the other hand, the WLS approximation needs only one unstable simulation[11], i.e., more relaxed conditions than the triangle.

When two convenient unstable simulations exist, the triangle and WLS approximations generally furnish numerical results close to each other.

3.2.5   Note on sensitivity analysis by SIME

The sensitivity computation of § 3.2.2 is a particular case of linearized sensitivity analysis.

Sensitivity analysis by SIME relies on stability margins. Sensitivity issues receive a large number of applications; they are elaborated in Chapter 3. Let us merely set here the basic formulation of a linearized sensitivity analysis for the sake of the developments of §§ 3.4, 3.5.

---

[10] see Section 2 of Chapter 6. This is also the reason for calling it "weighted" least-squares. Actually, in the context of Preventive SIME developed here this is simply a "least-squares" approximation.
[11] as close as possible to the critically unstable one; remember, the OMIB of concern can only be identified on an unstable simulation which identifies the pattern of machines' separation into two groups

By definition, the sensitivity coefficient of margin η with respect to parameter p is expressed by

$$S_p^\eta \triangleq \frac{\Delta\eta}{\Delta p} = \frac{\eta(k) - \eta(k-1)}{p(k) - p(k-1)} \qquad (2.26)$$

where k denotes the k-th simulation.

Accordingly, the value of parameter p which cancels the margin is given by

$$p|_{\eta=0} = p(k) - \frac{\eta(k)}{S_p^\eta} . \qquad (2.27)$$

In the above expressions, parameter p could be the clearing time, the power level or any other quantity liable to influence the system stability.

3.3 Existence and range of stability margins

3.3.1 General description

Stability margins in general express "how far" from (in)stability a system is.

The question is: does a stability margin as defined by SIME cover the whole range of stable and unstable cases ? If not, why and what is the range of existing margins ?

To explore this key issue, first recall that in the sense of SIME a stability margin results from the definition of an OMIB and the application of EAC. Now, according to EAC, a stability margin is defined in the $P - \delta$ plane in terms of the accelerating and decelerating areas, A_{acc} and A_{dec} (eq. (2.13)) or, equivalently, of the integral of P_a over δ (eq. (2.14)). In other words, existence of a stability margin in the sense of EAC implies existence of A_{acc} and A_{dec}, or, equivalently, of δ_u.

Second, observe that, strictly speaking, an OMIB is defined on unstable simulations only, since its existence is directly related to the machines' separation and loss of synchronism.[12]

The above observations suggest that, in the sense of SIME:

- negative margins can be defined only as long as A_{dec} (or, equivalently, δ_u) exists, that is as long as P_e goes past P_m for some $t \geq t_e$;

- positive margins can be defined only for a small range of cases close to the stability limit.

Taking as reference the limit stability conditions where $\eta = 0$, we see that in the sense of SIME, stability margins range from small positive values of η

---

[12] However, by continuation, we accept this OMIB to be valid on a stable simulation too, provided that this simulation is close to the borderline (un)stable case.

50   TRANSIENT STABILITY OF POWER SYSTEMS

up to negative values bounded by $|\eta_{\max}|$ which corresponds to $P_{a\min} = 0$ where

$$P_{a\min} = \min\{P_a(t) = P_m(t) - P_e(t), t > t_e, P_a(t) > 0\}. \quad (2.28)$$

Figure 2.4a illustrates this definition (see also Fig. A.5). In turn, $P_{a\min}$ depends on the contingency clearing time, t_e, under fixed operating conditions or, on P_m under fixed contingency clearing time. Below we explore on an example the range of existence of η.

3.3.2   Illustrations

Let us search on the 3-machine system the extremum of t_e under fixed P_m, and the extremum of P_m under fixed t_e.

Extremum of t_e. Let us consider the two stability cases described in Figs 2.1 and 2.2. They were carried out for respectively, $t_e = 117$ ms (for which $\delta_u = 158.0°$ (unstable simulation)), and $t_e = 92$ ms (for which $\delta_r = 146.7°$ (stable simulation)). Obviously, the borderline clearing time, i.e., the critical clearing time, CCT, lies in between 117 and 92 ms. It was also mentioned that for $t_e = t_c \equiv \text{CCT}$, $\delta_r = \delta_u = \delta_c$.[13] Angle δ_c is thus likely to be a little larger than δ_r and also larger than δ_u; indeed, decreasing t_e from 117 to CCT implies increasing A_{dec}, i.e., increasing δ_u.

Let us search $t_{e\min}$, i.e., the clearing time for which P_a starts being positive. Figs 2.4a, are drawn for various t_e's. They show that $t_e = 157$ ms yields an unstable case where a stability margin still exists, and that $t_e = 163$ ms is the borderline case, where the system stops having a post-fault equilibrium and hence unstable angle δ_u. Hence, $t_{e\min} = 163$ ms. Finally, $t_e = 190$ ms and, a fortiori $t_e = 250$ ms correspond to extremely severe stability conditions for which P_a has only positive values. For example, Fig. 2.4a shows that $P_{a\min}$ corresponding to $t_e = 250$ ms is 38.2 MW.

Extremum of P_m. Fig. 2.4b portrays a behaviour similar to that of Fig. 2.4a but for fixed clearing time, t_e, and variable OMIB mechanical power. One can see that the system stops having a post-fault equilibrium and corresponding unstable angle δ_u when P_m exceeds 149 MW.

3.3.3   Variation of salient parameters with t_e

Figures 2.5 illustrate the ranges of variation with t_e of $P_{a\min}$, η and $\delta_u(\delta_r)$, for the 3-machine system and operating conditions of Table 2.1, § 4.2. These figures are further commented below.

---

[13] Recall that subscript c stands for "critical" or limit stability conditions.

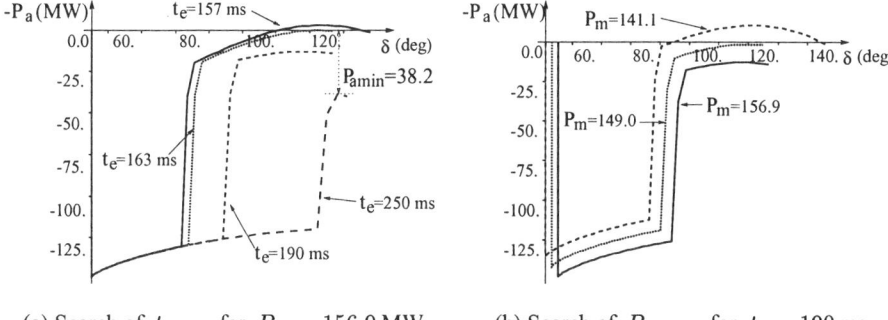

(a) Search of $t_{e\,min}$ for $P_m = 156.9$ MW (b) Search of $P_{m\,max}$ for $t_e = 190$ ms

Figure 2.4. $P_a - \delta$ representations for overly unstable cases. 3-machine system. Contingency Nr 2

Variation of $P_{a\,min}$ with t_e. The "distance" $P_{a\,min}$ defined by (2.28) is expressed in MW. Its variation is portrayed in Fig. 2.5a: it is lowerbounded by $t_{e\,min}$, the minimum value of t_e for which P_a becomes positive ($t_{e\,min} = 163$ ms) then starts increasing: its range goes from $t_{e\,min}$ ($\gg t_c$) onwards. Note the (almost) linear behaviour of the curve.

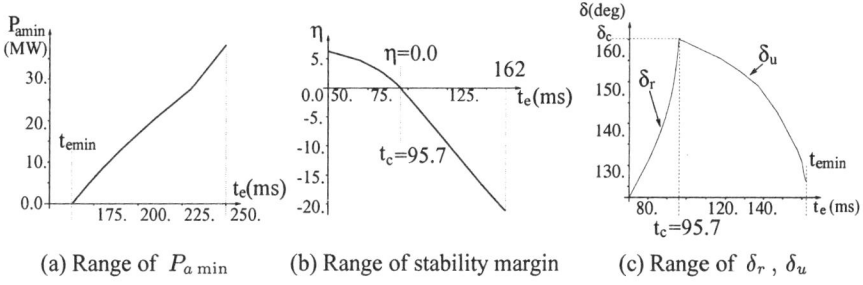

(a) Range of $P_{a\,min}$ (b) Range of stability margin (c) Range of δ_r, δ_u

Figure 2.5. Range of existence of $P_{a\,min}$, η, δ_r, δ_u. Contingency Nr 2

Variation of η with t_e. The curve η vs t_e is lowerbounded by a value of t_e smaller than but close to t_c and upperbounded by $t_e = t_{e\,min}$. Figure 2.5b portrays this variation. Note that the curve is almost linear for negative values, and also linear for $t_e \geq 75$ ms. For smaller values, the validity of the stability margin becomes doubtful. The margin becomes zero at 95.7 ms; hence, CCT$\equiv t_c = 95.7$ ms.

Variation of δ_r and δ_u with t_e. As already discussed, the curve δ_r vs t_e reaches a maximum at $t_e =$ CCT $= t_c$; its validity is restricted to the vicinity of t_c: $t_e \leq t_c$. The curve δ_u vs t_e is defined in the interval $(t_c, t_{e\,min})$.

52   TRANSIENT STABILITY OF POWER SYSTEMS

Figure 2.5c portrays these curves. We can also check that for the limit condition $t_e = t_c =$CCT, the three angles coincide: $\delta_r = \delta_u = \delta_c$.

3.4   A convenient substitute for unstable margins

It was just observed that under very stringent stability conditions the EAC margin defined by eqs (2.13), (2.14) does not exist anymore. This may arise for the following two reasons:

(i) there is no intersection of P_m and P_e curves (see above §§ 3.3.1 to 3.3.3)[14]
(ii) the clearing angle, δ_e, is beyond the unstable angle δ_u.

Hence, since all SIME-based computations rely essentially on stability margins, it is important to decrease the severity of the stability conditions so as to "effectively" reach the range of margins' existence.

Such a systematic procedure is derived below. It relies on a sensitivity computation similar to that of § 3.2.5, where

- the margin η is replaced by $P_{a\,\min}$ expressed by (2.28)
- the parameter p stands either for P_m, the OMIB mechanical power, or for t_e, the contingency clearing time.

(i) To appraise the amount of OMIB's mechanical power necessary to reach δ_u (where $P_a = 0$, see Fig. 2.6a), the sensitivity computation consists of writing:

$$P_m = P_m(k) - \frac{P_{a\,\min}(k)}{S_{P_m}^{P_{a\,\min}}} \quad (2.29)$$

with

$$S_{P_m}^{P_{a\,\min}} \triangleq \frac{\Delta P_{a\,\min}}{\Delta P_m} = \frac{P_{a\,\min}(k) - P_{a\,\min}(k-1)}{P_m(k) - P_m(k-1)}. \quad (2.30)$$

(ii) Similarly, to appraise the clearing time for which the OMIB meets δ_u (see Fig. 2.6b), the sensitivity calculation consists of writing:

$$t_e = t_e(k) - \frac{P_{a\,\min}(k)}{S_{t_e}^{P_{a\,\min}}} \quad (2.31)$$

with

$$S_{t_e}^{P_{a\,\min}} \triangleq \frac{\Delta P_{a\,\min}}{\Delta t_e} = \frac{P_{a\,\min}(k) - P_{a\,\min}(k-1)}{t_e(k) - t_e(k-1)}. \quad (2.32)$$

---

[14]This case corresponds to the absence of solutions of the dynamic eq. (2.2), i.e., of equilibria in the system post-fault configuration.

Chapter 2 - INTRODUCTION TO SIME 53

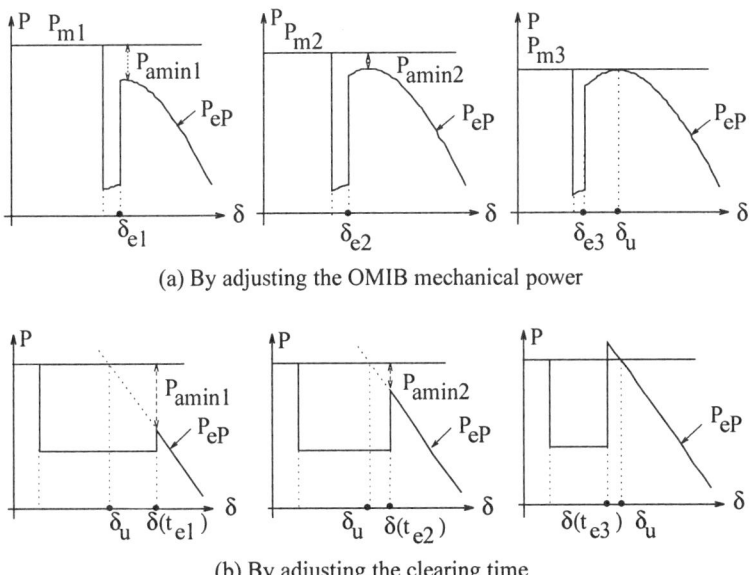

Figure 2.6. Procedures for reaching the range of margins' existence. Adapted from [Zhang et al., 1997a]

In both cases, this yields an iterative procedure which usually requires 2 to 4 iterations. Note that the sensitivity computation per se is virtually inexpensive. Hence, the computing time virtually reduces to the time spent for the T-D simulations; and since the cases of concern are generally very unstable, this time is generally short.

Figures 2.6a illustrate the procedure corresponding to above case (i), and Figs 2.6b to that of case (ii).

3.5 Next candidate CMs and margins

[15] So far, we have focused on the critical OMIB and resulting margin. In many applications, however, it is interesting to explore candidate OMIBs next to the actual critical OMIB. Their usefulness will appear in the following chapters. Let us only mention here that in various sensitivity-like applications one is led to consider two successive simulations (say, $(k-1)$ and k) and compare their corresponding margins. Now such a comparison is possible only if the simulations refer to the same set of CMs and resulting OMIB. If not, one should continue running simulations until getting two successive simulations with identical OMIBs.

---

[15]This paragraph deals with matters aiming at speeding up SIME's algorithm. It might be skipped at first reading.

54   TRANSIENT STABILITY OF POWER SYSTEMS

A convenient solution may be found in the use of candidate modes of separation (and corresponding candidate OMIBs) close to (though not identical with) the actual mode of separation. Such situations often arise on large power systems having hundreds of machines for which the mode of separation can be rather versatile, leading to two successive simulations[16] which, even close, do not have exactly the same CMs. The solution then consists of comparing the margins of simulations k and $(k-1)$ computed for the OMIB of simulation k.

To state this more clearly, let us denote COMIB a candidate OMIB and $\eta(\text{COMIB})$ the corresponding margin. Then, if $\text{OMIB}(k)$ is identical to $\text{COMIB}(k-1)$, we may, for example, extrapolate the margins $\eta(\text{OMIB}(k))$ and $\eta(\text{COMIB}(k-1))$. Obviously, the result won't be "optimal"; nevertheless, it is likely to be good enough to allow saving one T-D simulation.

ILLUSTRATION. The 3-machine system considered so far has limited possibilities. This is why the example used below and illustrated on Figs 2.7 does not conform to the above scheme. Rather, it consists of computing the margins $\eta(k), \eta(k-1)$ corresponding to the clearing times $t_e = 177$ and 137 ms respectively, for the candidate OMIB composed of one machine (m_2). The reason for using the candidate instead of the actual critical OMIB is that this latter does not have margin for $t_e = 177$ ms, while for the candidate OMIB it yields: $\eta(177) = -28.25$; $\eta(137) = -12.64$. Their linear extrapolation provides an approximate CCT of 105 ms. (Remember, CCT corresponds to zero margin.)

Given that the actual CCT is 95.7 ms, we see that the above approximation is not so bad, after all. Figures 2.7 illustrate the procedure.

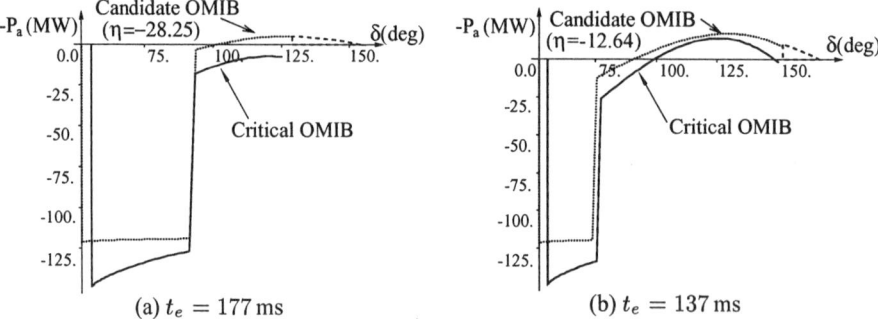

Figure 2.7. $P - \delta$ representations of the critical and the next to the critical OMIBs. 3-machine system. Contingency Nr 2

---

[16] Successive simulations are for example simulations run for slightly different clearing times or operating states.

3.6 Normalized margins

According to eq. (2.14), the margins considered so far are expressed in MW.rad. On the other hand, a closer look at the expression of the negative margin (2.19) shows that its numerical value is proportional to the OMIB inertia coefficient. In turn, this coefficient depends on the number of system machines, their inertia and their mode of separation (i.e., the way the machines separate into CMs and NMs).

To make the range of margin values less dependent on the above factors, we use "normalized" margins by dividing expressions (2.19), (2.21) by the OMIB inertia coefficient. These normalized margins are thus expressed in $(\text{rad/s})^2$.

Along with the same lines, the "normalized distance" $P_{a\min}$, is obtained by dividing $P_{a\min}$ (MW) by the OMIB inertia coefficient, and is expressed in rad/s^2.

Let us insist that, whether normalized or not, margins and "accelerating power distances" do not have the same meaning and cannot be compared to each other.

4. SIME'S TYPICAL REPRESENTATIONS

So far, SIME has illustrated various aspects of transient stability phenomena, via two OMIB representations: the time-domain and the $P - \delta$ ones. In this section a third representation will be introduced: the OMIB phase plane (or phase portrait).

These representations will be applied first on the 3-machine system, then on a real-world power system.

In these simulations, SIME's accuracy is assessed in terms of CCTs: CCT by SIME as compared with the CCT of the corresponding T-D program coupled with SIME. Note that the CCT computation by SIME is performed according to § 2.2 of Chapter 4.

4.1 Illustrations on the three-machine system

4.1.1 Stability conditions

The stability case consists of the 3-machine system under the prefault operating conditions and the contingency considered so far (Nr 2).

The contingency is cleared at different clearing times, t_e's, varying in between 250 ms and 70 ms, i.e., above and below the critical clearing time, t_c, which was found to be 95.7 ms. In all these simulations, the critical separation pattern was found to split the machines into two CMs (m_2 and m_3) and one NM (m_1).

4.1.2 OMIB parameters and numerical results

Table 2.1 gathers the 3-machine system parameters, under the operating conditions considered so far and contingency Nr 2, for $t_e = 162\,\mathrm{ms}$. It also lists the OMIB parameters, computed according to eqs (2.3) to (2.12). Note that with the modeling considered here, the machine mechanical powers P_{m_1}, P_{m_2}, P_{m_3} are constant, and so is the OMIB mechanical power P_m, computed via eqs (2.5), (2.9).

Table 2.1. 3-machine system and OMIB parameters; $t_e = 162\,\mathrm{ms}$

P_{m_1}	72
P_{m_2}	163
P_{m_3}	85
M_1	12.54
M_2	3.39
M_3	1.59
M_C	4.99
M_{NC}	12.54
M	3.57
P_C	177.40
P_{NC}	20.49
P_m	156.90
η	-75.20
δ_u	126.3°

Table 2.2. Stability conditions vs t_e

t_e (ms)	$\eta\,(P_{a\,\min})$ not normal.	$t_u\,(t_r)$ (ms)
250	(-38.28)	261
210	(-20.64)	241
190	(-12.96)	261
177	(-7.14)	263
167	(-2.25)	263
163	(-0.21)	274
162	-75.20	273
157	-70.17	308
137	-48.01	373
117	-24.50	458
96	-0.22	773
95	1.37	(827)
92	3.47	(663)
89	8.42	(615)
80	5.47	(542)
70	10.11	(512)

On the other hand, Table 2.2 gathers information relative to various t_e's for the stability conditions of Table 2.1. The values of η or $P_{a\,\min}$ (listed between brackets) are not normalized. The values of $t_u\,(t_r)$ are in ms. Note that the listed positive margins are provided by the triangle approximation.

4.1.3 SIME's three representations

The OMIB swing curves, $P - \delta$ plots and phase plane representations are displayed in Figs 2.8, under various contingency clearing times. More precisely, Figs 2.8a and 2.8b display the swing curves and $P - \delta$ curves for respectively $t_e = 177\,\mathrm{ms}$ and $t_e = 137\,\mathrm{ms}$; Fig. 2.8c plots the phase plane representation corresponding to four clearing times: 177 ms, 137 ms, 117 ms (for which the swing and $P - \delta$ curves are portrayed in Figs 2.1) and 92 ms (see Figs 2.2).

Chapter 2 - INTRODUCTION TO SIME 57

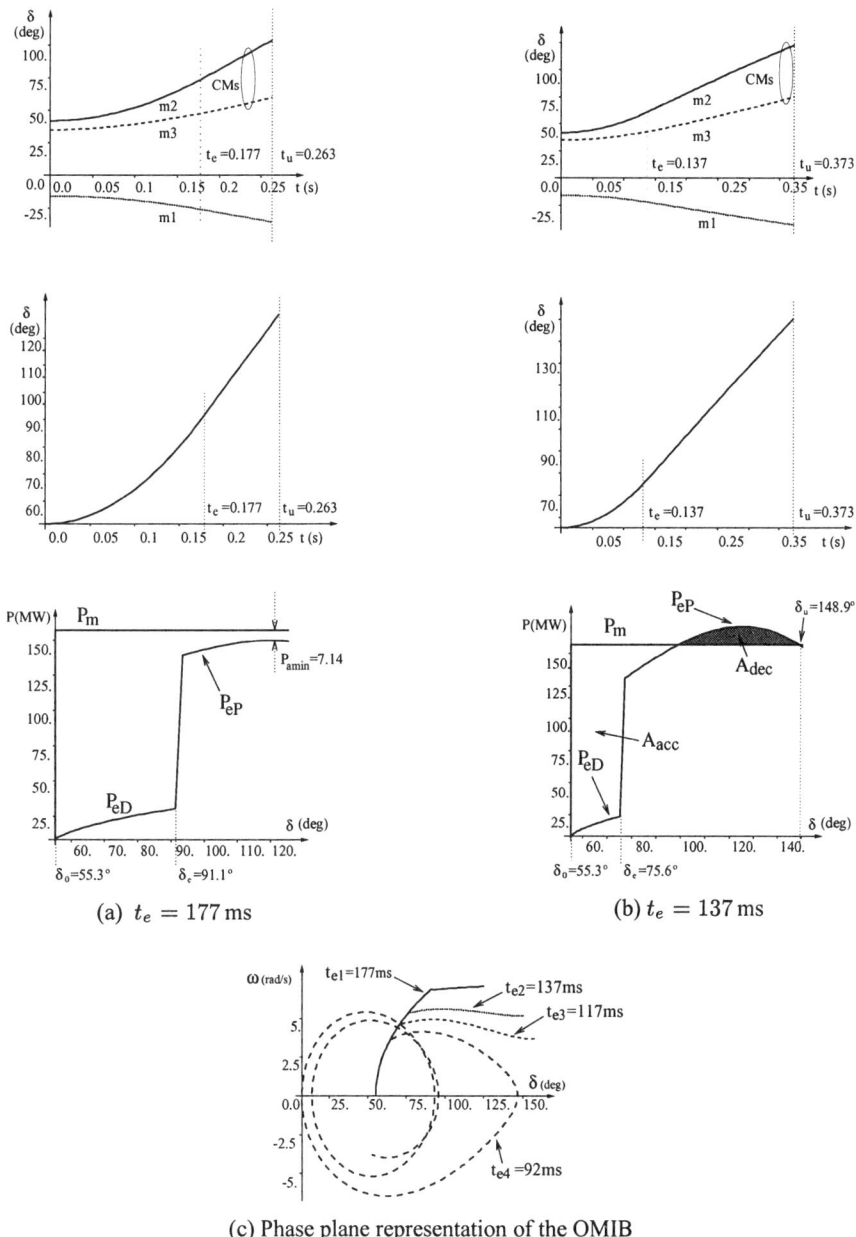

(a) $t_e = 177$ ms

(b) $t_e = 137$ ms

(c) Phase plane representation of the OMIB

Figure 2.8. Three SIME's representations for the 3-machine system. Contingency Nr 2. CCT(SIME) = 95.7 ms ; CCT(MATLAB) = 95 ms

58 TRANSIENT STABILITY OF POWER SYSTEMS

Discussion. Obviously, the OMIB swing curve is an image of the multimachine system time evolution. In other words, the information about the multimachine time evolution is compressed into that of the OMIB evolution, with a compression rate of n, the number of system machines.

Of course, in this oversimplified academic example the compression rate is low. And since the multimachine swing curves are quite smooth, the OMIB swing curves do not add much to the description of the phenomena.

One might therefore think that these curves do not contribute significantly to understand and/or interpret the multimachine stability phenomena. This, however, is not true anymore when it comes to real power systems with intricate stability behaviour, like those discussed below, in § 4.2.

The EAC $P - \delta$ representation, on the other hand, allows getting a different picture of the phenomena. Its simplicity enhances their understanding; besides, it opens avenues to an impressive number of applications of all three types: analysis, sensitivity analysis and control (i.e., stabilization). The resulting advantages are paramount, as will be seen in the remaining chapters.

Finally, the OMIB phase plane representation provides a typical system theory description of transient stability phenomena, conveying complementary type of information. Obviously, clearing times t_{e1}, t_{e2}, t_{e3} drive the system to instability, since the corresponding trajectories escape from the stability region; on the contrary, the trajectory corresponding to t_{e4} spirals inwards, toward the stable post-fault equilibrium solution.

It is interesting to observe that the information contained in the phase plane results from a compression rate of $N/2$, where N denotes the total number of system state variables. For example, in the present 3-machine system, N equals 21 (2 mechanical, 2 electrical and 3 AVR's state variables per machine). Hence, the compression rate is 21/2. This rate is impressive even for this academic example. It becomes dramatic in real-world cases, while, at the same time, preserving the clarity of interpretations. This is illustrated on the following example.[17]

4.2   Illustrations on the Hydro-Québec system

Main characteristics of the Hydro-Québec system are provided in Section 2 of Appendix B. Let us only specify here that it is represented by 86 machines, and that these machines together with the network total over 2,000 state variables.

The contingency considered here creates back- and multi-swing phenomena, that we will further analyse in Section 5.

Figures 2.9 and 2.10 illustrate these phenomena in the three SIME's representations: swing curves, $P - \delta$ curves, phase plane curves.

---

[17] See also the example of § 2.2.6 of Chapter 4, and corresponding Figs 4.3, 4.6 and the example of § 1.4 of Chapter 7.

Figures 2.9a correspond to an unstable case ($t_e = 198$ ms), Figs 2.9b to a stabilized case ($t_e = 175$ ms).

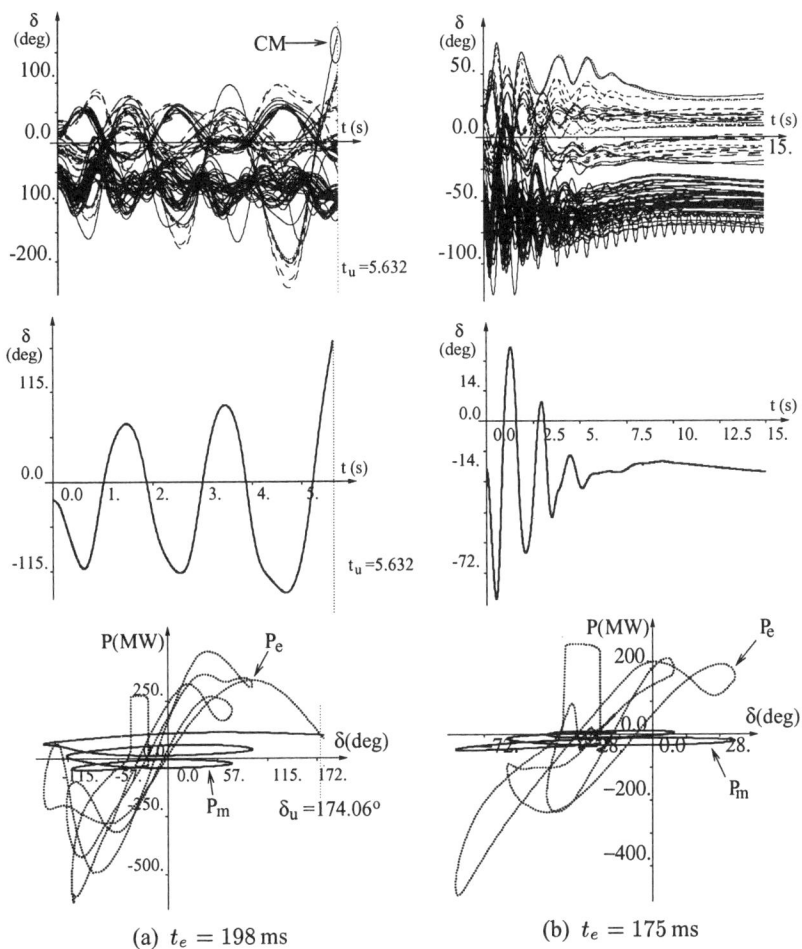

Figure 2.9. Illustration of back- and multi-swing phenomena on the Hydro-Québec system: swing and $P - \delta$ curves. CCT(SIME) = 175.5 ms ; CCT(ST-600) = 176 ms.

Incidentally, note that the multiswing CCT is slightly larger than 175 ms. Further, note that the first-swing CCT equals 220 ms, i.e., it is fairly larger than the multiswing one. But the system goes first-swing unstable after 418 ms, i.e., much earlier than multiswing unstable (5.63 s). In other words, for t_e between 175 and 220 ms the system will lose synchronism after a few seconds, while for t_e larger than 220 ms it will lose synchronism after some hundred milliseconds.

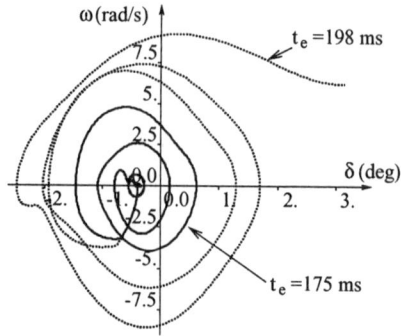

Figure 2.10. Phase plane representation of back- and multi-swing phenomena of Figs 2.9

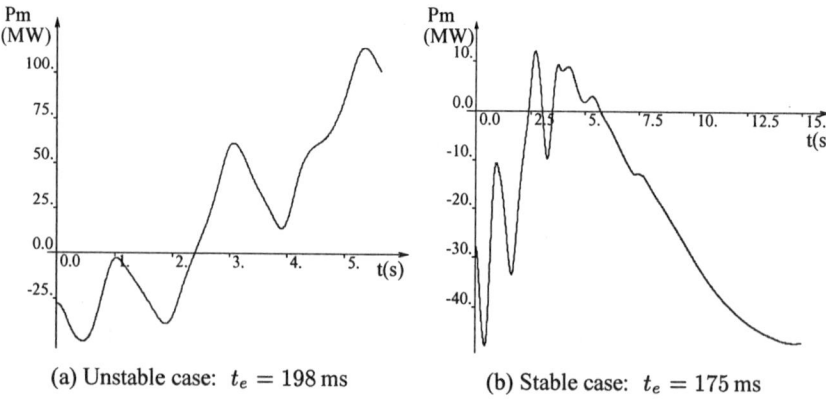

Figure 2.11. Time evolution of the OMIB mechanical power of the Hydro-Québec system

According to the discussion of § 4.1.3, the compression rate of the OMIB swing curves is 86, while that of the phase plane is over $2,000/2 = 1,000$.

Even more important than this dramatic compression rate is that these OMIB representations give an insight into the multimachine system behavior.

Indeed, the OMIB swing curves appraise phenomena hardly possible to observe on the multimachine swing curves, because of their high intricacy.

On the other hand, the phase plane of Fig. 2.10 describes simply and clearly the backswing and multiswing phenomena: multiswing instabilities, since the unstable trajectory spirals outwards the stable equilibrium, before definitely escaping from the stability region; backswing (stable and unstable) phenomena, since the trajectories start spiraling backwards, contrary to those of Fig. 2.8c.

Finally, observe that the pre-fault and post-fault stable equilibrium points are slightly different, suggesting a small change in the topology.

Coming back to the $P-\delta$ plots of Fig. 2.9, observe the important OMIB's P_m vs δ variations, due to the system turbines' governors. More precisely, observe that at the first instants of the fault application, the mechanical power of the OMIB decreases, despite its deceleration. One may wonder why the machines' mechanical power regulation tends to amplify this instability phenomenon. Actually, this is caused by the hydraulic type of system turbines, which are non-minimum phase. Hence, at the very first instants, although their primary regulation aims at increasing their mechanical power, it actually decreases it. For information, Figs 2.11 plot the corresponding evolution of P_m with time.

4.3 SIME as a reduction technique

The above three representations uncover an interesting interpretation of SIME: that of a technique allowing a dramatic reduction of the dimensionality of the power system transient stability problem. This significant result provides SIME with extraordinary flexibility, strength and possibilities.

5. BACK- AND MULTI-SWING PHENOMENA

5.1 Definitions

Multiswing phenomena along with the number of corresponding swings receive a clear interpretation and assessment through the OMIB representations, like those of Figs 2.9, 2.10.

Observe that in Fig. 2.9a the OMIB is composed of 1 CM and 85 NMs. Observe also that the CM and the OMIB start in a backswing (decelerating) mode; but after some oscillations, they lose synchronism in an upswing (accelerating) mode. On the other hand, Figs 2.9b are relative to a stabilized case.

The OMIB swing curve provides a clear insight and interpretation of the physical phenomena and readily yields the following definitions.

An m-swing (in)stability $(m > 1)$ appears if the OMIB angle during the $(m - 1)$ swings does not meet its unstable angle. Thus, the m-swing will be

- unstable if the OMIB reaches its unstable angle, defined by conditions (2.15); for example, for the unstable case of Fig. 2.9a, $m = 3$

- stable if the OMIB reaches its return angle defined by conditions (2.16).

A final observation: the critical machines' identification procedure of § 2.1 applies to any swing curve; it is therefore valid for multiswing phenomena also.

5.2 Analytical expression of margins

The general first-swing margin expression (2.14) holds valid for multiswing phenomena as well, through proper adjustment. Thus,

$$\eta = -\int_{\delta'_0}^{\delta'_u} P_a \, d\delta \qquad (2.33)$$

where δ'_0, δ'_u stand respectively for the initial and unstable OMIB angles of the m th swing. The unstable angle is defined as for the first swing, by conditions (2.15), except that here t_0 stands for the time taken to reach δ'_0.[18]

Like first-swing stability, eq. (2.33) yields the analytical expressions of multiswing stable and unstable margins as

$$\eta_{st} \simeq \frac{1}{2}|P'_{ar}|(\delta'_u - \delta'_r) \qquad (2.34)$$

$$\eta_u = -\frac{1}{2}M\omega'^2_u \qquad (2.35)$$

where P'_{ar}, δ'_r, δ'_u, ω'_u stand, respectively, for accelerating power, return angle, unstable angle and unstable speed of the OMIB of the appropriate swing number.

6. DIRECT PRODUCTS AND MAIN BY-PRODUCTS
6.1 Description

Broadly, one may distinguish three types of information provided by SIME: stability margins and critical machines (CMs); EAC per se; sensitivity analysis and control techniques.

Stability margins and CMs are the two major direct achievements of SIME. This chapter has essentially focused on their formulation, derivation and exploration.

The second type of SIME's source of information is the very EAC, applied to OMIB, i.e. to a faithful image of the multimachine system behaviour under given modeling. The sheer number of physical interpretations that EAC provides to intricate transient stability phenomena is certainly not the least asset of the method.

The third type of SIME's information, obtained as a by-product of the previous two, are sensitivity analysis and control.

Sensitivity analysis with respect to a given parameter relies on the existence of stability margins and deals with their variation with this parameter. Although sensitivity coefficients may be assessed only numerically, their use is

---

[18] The initial angle of the m th swing is the OMIB's minimum angular deviation reached at the end of the $(m-1)$ th swing.

generally very simple and straightforward. Sensitivity analysis is a powerful multiobjective tool scrutinized in Chapter 3.

On the other hand, control relies on the combined use of margins, CMs, and suggestions provided by EAC. Control techniques devise means to determine the type and size of action necessary to stabilize an otherwise unstable scenario, be it in the preventive or the emergency mode. Control techniques are introduced in Chapter 3 and elaborated in Chapter 4.

Sensitivity analysis and control techniques together with stability margins, CMs, and the very EAC will allow us to tackle any type of transient stability study, i.e., to cover the whole field of transient stability assessment and control (TSA&C).

6.2 Organization of topics

The remainder of this monograph is devoted to the development of transient stability techniques of the preventive and the emergency modes.

Chapter 3 focuses on sensitivity analysis in general, and the derivation of various techniques on which rely the various types of applications derived in the following chapters.

Chapters 4 and 5 cover all aspects of TSA&C in the preventive mode.

Chapter 6 introduces a new technique: the real-time closed-loop emergency control.

Finally, Chapter 7 overviews salient features of preventive and emergency SIMEs and gives a synthetic comparison of the three classes of existing approaches to transient stability.

7. PREVENTIVE vs EMERGENCY SIME

Differences between Preventive and Emergency SIME were shortly mentioned in §§ 2.3, 2.4 of Chapter 1 and in § 1.2 of this Chapter. They will fully be specified in the following chapters. However, as a convenience to the reader, below we summarize their basic essentials.

In essence, SIME combines information about the multimachine power system with the OMIB stability assessment based on EAC.

The multimachine information is obtained either from transient stability simulations of contingencies likely to occur, or from measurements acquired from the system power plants in real-time and reflecting the actual occurrence of a contingency. The former information is processed by the Preventive SIME, the latter by the Emergency SIME.

Both SIMEs aim at performing successively two main tasks: transient stability assessment and control. But while the Emergency SIME attempts to control the system *just after* a contingency occurrence and its clearance, so as to maintain synchronism, the Preventive SIME aims at proposing countermeasures *preventively*, i.e. *before* the actual occurrence of any contingency.

7.1 Preventive transient stability assessment

Preventive TSA goes along the traditional way of assessing the system robustness vis-à-vis occurrence of anticipated contingencies. In an on-line context, preventive TSA should consider all plausible contingencies, in a time horizon of, say, 30 minutes ahead: computational efficiency becomes thus crucial.

On-line preventive TSA may effectively be decomposed into contingency filtering (to detect existence of harmful contingencies while discarding the (large majority of) harmless ones), contingency ranking (to classify the contingencies found to be potentially harmful, according to their degree of severity), and contingency assessment, to scrutinize the harmful ones.

According to the Preventive SIME, these tasks are achieved using T-D simulations to determine stability margins and critical machines. The resulting filtering and ranking procedures are significantly faster than those relying on conventional pure T-D approaches. Nevertheless, the paramount advantage is control.

7.2 Predictive transient stability assessment

Unlike preventive TSA, predictive TSA has not been used so far, for two main reasons: on one hand, because this task is hardly achievable - if at all - by conventional approaches; on the other hand, because its interest is directly linked to the feasibility of closed-loop emergency control and, again, this cannot be tackled by conventional approaches.

Unlike preventive TSA, the predictive TSA deals, in real-time, with (succession of) events which have been detected but not necessarily identified, and generally automatically cleared by the protective devices. Thus, in order to be effective, the *prediction* of the system behaviour must be performed *early enough* so as to leave sufficient time for determining and triggering appropriate control actions, whenever necessary. To get a stability diagnostic ahead of time, the predictive TSA relies on real-time measurements (see Chapter 6).

7.3 Control

Whether for preventive or for emergency use, the very design of control techniques relies on common principles. But the way of applying them differs substantially in many respects (speed of action, closed-loop vs open-loop fashion, character of absolute necessity or not, etc.).

8. SUMMARY

SIME's roots have been traced back to the Lyapunov's direct method, and its features as an OMIB-based method were specified. It was suggested that the conjunction of direct criteria with detailed temporal information about multimachine parameters (provided by a T-D program or by real-time measurements) allows evading the weaknesses of both (direct criteria and temporal information) while keeping their strengths and even magnifying them substantially.

It was pointed out that the basic information furnished by SIME concerns stability margins and critical machines. This chapter has essentially focused on their derivation and specifics.

The results were illustrated mainly on a three-machine example to help the reader get familiar with essential aspects; on the other hand, real-world stability cases were used to uncover phenomena specific to large-scale systems.

Through these developments and illustrations it was shown that SIME may be seen as a reduction technique able to transform the multidimensional state space of the multimachine power system into a two-dimensional phase plane. The resulting plot is a faithful picture of the multimachine stability phenomena, no matter whether they correspond to a small test system, modeled in a rather simplified way, or to a large-scale real system modeled in detail.

Chapter 3

SENSITIVITY ANALYSIS

Chapter 2 has dealt with two major achievements of SIME: calculation of stability margins and identification of critical machines; stability margins are at the very core of SIME-based sensitivity analysis.

In a broad sense, sensitivity may be viewed as the ratio of response to cause and appraised in terms of sensitivity coefficients. In the context of control theory, sensitivity coefficients express the dependence of a system state variable or a performance index on system parameters.

In transient stability studies, "the performance index" should naturally be related to (the degree of) system stability. The margin provided by SIME appears to be a very powerful stability index. As for parameters likely to have significant impact on power system stability, we may distinguish "elementary" and "global" ones. Elementary parameters can for example be: contingency characteristics (e.g., clearing scenario) or network characteristics (e.g., number of lines on a heavily loaded corridor, FACTS devices, etc). Global parameters, on the other hand, are meant to be suitable combinations of elementary ones; OMIB parameters developed in the context of SIME are good candidates.

Like in many other fields, an ultimate goal of transient stability sensitivity analysis is control, i.e., design of corrective actions able to stabilize an otherwise unstable case. Now, in the sense of SIME, this may be viewed as the direct extension of sensitivity investigations, since the purpose is to cancel out the corresponding negative margin.

This chapter lays the foundations of sensitivity analysis techniques. More precisely, the objectives are:

- *to describe sensitivity analysis in general, scrutinize conventional methods, and show that the use of SIME's margin makes sensitivity analysis a valuable tool in transient stability studies;*

- to derive SIME-based sensitivity analysis;
- to devise SIME-based compensation schemes which provide one-shot procedures for calculating stability limits ;
- in short, to lay the foundations of key techniques for TSA&C.

1. ELEMENTS OF SENSITIVITY ANALYSIS
1.1 Problem statement

The sensitivity of a system to variations of its parameters plays an important role in analysis and control. Indeed, on one hand, the values of a system parameters differ from the nominal ones (because of inaccuracies in the computed data, of their variation with time, of imperfect realization of the controlled devices). On the other hand, information about the way the characteristics of a system depend on its parameter variations may be used to improve its performances as, for example, in adaptive systems.

Power system dynamics in general is governed by the sets (1.1), (1.2) of differential-algebraic equations re-written below as:

$$\dot{x} = f(x, y, p) \quad (3.1)$$
$$0 = g(x, y, p) \quad (3.2)$$

where, x is the n-dimensional state vector, y the m-dimensional vector of algebraic variables, and p the k-dimensional vector of parameter variables. The m algebraic equations define a manifold of dimension $n + k$, the *constraint manifold*, in the $(n + m + k)$-dimensional space of x, y and p.

Differential-algebraic systems can be analyzed using the implicit function theorem. According to this theorem, under the assumption that the algebraic Jacobian $g_y(x, y, p)$ is non-singular at point x, y and p, there exists a locally unique, smooth function F of the form:

$$\dot{x} = f(x, h(x, p), p) = F(x, p) \quad (3.3)$$

from which the algebraic variables have been eliminated. Equations (3.3) are preferred to the set (3.1), (3.2) because they get rid of any constraint manifold.

Sensitivity analysis can be related to various types of variations : those which do not alter the order of the system or its initial conditions are referred to as α-variations; variations of the initial conditions, called β-variations; variations leading to changes in the system order, the λ-variations. For example the change of a feedback gain can be referred to as an α-variation; a change of the fault duration can be considered as a β-variation if the state vector of the system entering its post-fault configuration is considered as the initial state. Power rescheduling among a group of generators can be considered as a combination of both β-variations (changes in machine generation modify the state space of the system) and α-variations (generation power references are modified). On the other hand, λ-variations are mostly due to the presence of small parameters not taken into account in the mathematical system modeling; for example, they could be introduced by the placement of a new control device which increases the system order [Kokotovic and Rutman, 1965].

Three types of sensitivity analysis methods are briefly described below: (i) *sensitivity analysis of the linearized system*; (ii) *analysis of supplementary motion* which studies the influence of parameter variations on the system motion; (iii) sensitivity analysis based on performance indexes rather than system state variables.

N.B. Paragraphs 1.2 to 1.5 of this section deal with theoretical aspects of conventional sensitivity analysis methods, and attempt to highlight, by comparison, the significance of SIME's contributions in the field. The reader who is familiar with such methods may skip these paragraphs and go directly to § 1.6.

1.2 Sensitivity analysis of the linearized system

The linearization of the Differential-Algebraic (D-A) system described by eqs (3.1) and (3.2) leads to:

$$\begin{bmatrix} \Delta \dot{x} \\ 0 \end{bmatrix} = J \begin{bmatrix} \Delta x \\ \Delta y \end{bmatrix} \tag{3.4}$$

where J is the *unreduced* Jacobian of the D-A system:

$$J = \begin{bmatrix} f_x & f_y \\ g_x & g_y \end{bmatrix}. \tag{3.5}$$

For complex systems, where the Jacobian cannot be calculated analytically, a numerical method is used. Starting from the states determined from model initialization, a small perturbation is applied to each state successively. The variation of all the states divided by the magnitude of the perturbation gives a column of the Jacobian matrix corresponding to the disturbed state.

Assuming that g_y is nonsingular we eliminate Δy from eq. (3.4) and get:

$$\Delta \dot{x} = [f_x - f_y g_y^{-1} g_x] \Delta x = A \Delta x. \tag{3.6}$$

In the power system literature, the state matrix A of the linearized system is often called the *reduced* Jacobian as opposed to the unreduced one. This matrix, which depends on vector p can also be viewed as the result of linearization of eq. (3.3).

70   TRANSIENT STABILITY OF POWER SYSTEMS

The sensitivity analysis of the linearized system, via matrix A, may be performed using standard techniques of linear systems sensitivity analysis. Sensitivity techniques applied to linearized systems are efficient even for very large-scale systems. But their validity depends on the validity of the system linearization, which is questionable for highly non-linear phenomena as those involved in transient stability.

1.3   Sensitivity analysis of the supplementary motion

The original system (3.3)[1] will be decomposed here into an equation for each one of its i-th components:

$$\dot{x}_i = F_i(x_1, ..., x_n; p_1, ..., p_k) \qquad (i = 1, ..., n) \qquad (3.7)$$

with initial conditions $x_i(0) = x_i^0$.

If we consider that the parameter values undergo variations Δp_j, then the varied motion is described by the system[2]

$$\dot{\tilde{x}}_i = F_i(\tilde{x}_1, ..., \tilde{x}_n; p_1 + \Delta p_1, ..., p_k + \Delta p_k),$$
$$\tilde{x}_i(0) = x_i^0 \qquad (i = 1, ..., n). \qquad (3.8)$$

System (3.8) corresponds to the varied motion $\tilde{x}_i(t)$, while system (3.7) to the fundamental or unperturbed motion. The *supplementary motion* is defined as the difference

$$\Delta x_i(t) = \tilde{x}_i(t) - x_i(t) \qquad (i = 1, ..., n). \qquad (3.9)$$

Assuming that the solutions of (3.7) and (3.8) are differentiable with respect to p_j, we represent the supplementary motion (3.9) in a Taylor series expansion in powers of p_j. If we can accept to limit the expansion to the first linear terms, we get:

$$\Delta x_i(t, \Delta p_1, ..., \Delta p_k) = \sum_{j=1}^{k} (\frac{\partial x_i(t)}{\partial p_j}) \Delta p_j + ..., \qquad (3.10)$$

where the higher order variations of p_j will be considered to be negligible.

Thus, according to eq. (3.10), the study of supplementary motion leads to the calculation and analysis of partial derivatives

$$S_{p_j}^{x_i}(t) = \frac{\partial x_i(t)}{\partial p_j}\bigg|_{\Delta p_1 = 0, ..., \Delta p_k = 0}. \qquad (3.11)$$

---

[1] The original system can also be considered as the unperturbed system.
[2] except for β-variations.

The *sensitivity function* or *coefficient* $S_{p_j}^{x_i}(t)$ of the coordinate x_i relative to the parameter p_j, reflects the trend of variation of the coordinates under parameter variations. It is also known as trajectory sensitivity. Note that the sensitivity function can also be obtained as the solution of the m equations

$$\dot{S}_{p_j}^{x_i} = \sum_{k=1}^{n} \frac{\partial F_i}{\partial x_k} S_{p_j}^{x_k} + \frac{\partial F_i}{\partial p_j}. \qquad (3.12)$$

Note also that the partial derivative of $S_{p_j}^{x_i}(t)$ with respect to parameter p_j is given by:

$$\frac{\partial S_{p_j}^{x_i}(t)}{\partial p_j} = \frac{\partial^2 x_i(t)}{\partial^2 p_j}. \qquad (3.13)$$

It may be difficult or even impossible to get an analytical expression for the sensitivity function. We can then resort to numerical computations, relative to the temporal evolution of the original and the perturbed systems associated with the set of parameters $p_1, ..., p_j + \Delta p_j ..., p_k$. If Δp_j is small enough, then $S_{p_j}^{x_i}(t)$ can be suitably approximated by:

$$S_{p_j}^{x_i}(t) \approx \frac{\tilde{x}_i(t) - x_i(t)}{\Delta p_j}. \qquad (3.14)$$

Note that in the case of β-variations the varied motion corresponds to system (3.7) with different initial conditions $x_i(0) = x_i^0 + \Delta x_i^0$. However, if the initial conditions are also considered as system parameters, we can apply the same formulation. The formulation of varied motion for λ-variations is different and will not be discussed here.

Unlike sensitivity analysis of the linearized system, an analysis of the supplementary motion is able to reproduce correctly the phenomena involved in the original non-linear system, since it does not imply their linearization. Nevertheless, none of these two methods provide global information about the influence of parameter variations on the system behaviour.

This approach is also the basis of the research work reported in [Laufenberg and Pai, 1997, Hiskens et al., 1999, Hiskens and Pai, 2000].

1.4   Synthetic sensitivity functions (ssfs)

Such a global information may be obtained by using appropriate synthetic sensitivity functions (ssf for short). Such a function aims at expressing the dependence of a performance index on the corresponding set of parameters ($\boldsymbol{p}$ and $\tilde{\boldsymbol{p}}$) of the original and the perturbed systems.

Letting s be this function we write:

$$s(t) = ssf\left(\boldsymbol{x}(t), \tilde{\boldsymbol{x}}(t), \boldsymbol{p}, \tilde{\boldsymbol{p}}\right). \qquad (3.15)$$

Admittedly, expression (3.14) can also be considered as an ssf function. But in contrast to eq. (3.15), sensitivity analysis relying on system state variables can be an extremely tedious task, which, moreover, does not necessarily provide a comprehensive account of the physical phenomena.

For example, consider the influence of a parameter p_j on the dynamics of the state vector $x(t)$ when the system is subjected to a constant perturbation. An analysis of the supplementary motion would need the computation of n sensitivity functions $S_{p_j}^{x_i}(t)$, $i = 1, 2, ..., n$. And the influence of k parameters would imply $n*k$ sensitivity functions. Another even more serious difficulty comes from the fact that these sensitivity functions are time dependent; hence the necessity to analyze n trajectories. Note, however, that this difficulty might partly be circumvented by determining for each i the time $t_k \in [0, t_{end}]$ for which $|\tilde{x}_i(t) - x_i(t)|$ is maximum over the time interval and hence by considering as a synthetic sensitivity function, the expression:

$$S_{p_j}^{x_i} = \frac{\tilde{x}_i(t_k) - x_i(t_k)}{\tilde{p}_j - p_j}. \tag{3.16}$$

The existence of a convenient performance index yielding an appropriate synthetic sensitivity function may provide very interesting alternative solutions. Suppose for example that the output of the synthetic sensitivity function is a good picture of the parameters' influence on stability. Suppose further that $S_{p_j}^{\text{ssf}}$ is the scalar output of the ssf obtained with a perturbed motion relative to the parameter vector

$$\tilde{p} = [p_1, ..., p_j + \Delta p_j, \cdots, p_k]^T = p + [0, ..., \Delta p_j, ..., 0]^T.$$

Then the vector $[S_{p_1}^{\text{ssf}}, ..., S_{p_k}^{\text{ssf}}]^T$ gives in the k-dimensional parameter space the direction that provides the largest rate of the stability domain increase. Using this information, a steepest descent could determine the set of parameters corresponding to the largest stability domain.

Applications of this approach are described in [Rovnyak et al., 1997].

Below we give some simple illustrations on the 3-machine system.

1.5 Illustrative examples

1.5.1 Simulation conditions

We illustrate the above considerations by performing sensitivity analysis of transient stability phenomena on the three-machine system under the conditions described below.

This system is modeled with constant impedance loads and 7 state variables per machine: 2 mechanical variables (rotor angle and speed), 2 electrical variables and 3 variables for the AVR modeling.

The single-line diagram of this system in its prefault configuration is represented in Fig. 3.1a. Prior to any fault, i.e. for $t < 0$ the system is in

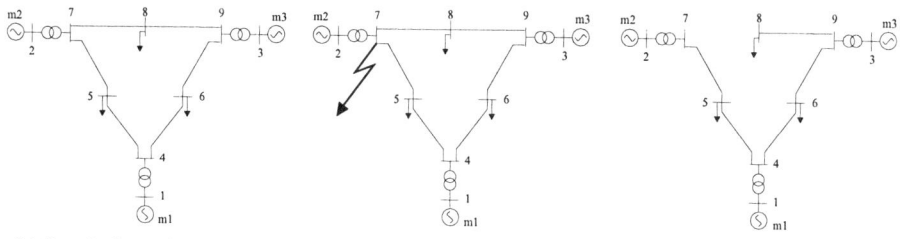

(a) Pre-fault configuration   (b) During-fault configuration   (a) Post-fault configuration

Figure 3.1.   One-line diagram of the 3-machine system

a steady-state operation; hence its state vector $\mathbf{x}(0)$ is a stable equilibrium solution of eqs (3.3). At time $t = 0$, the system suddenly changes configuration, because a three-phase short-circuit is applied at node #7 as illustrated in Fig. 3.1b. The duration of the fault is 150 ms: at $t = 150$ ms, the fault is cleared by opening the line connecting nodes #7 and #8, and the system enters its final configuration represented in Fig. 3.1c.

Since in transient stability machine angles and speeds are acknowledged to be variables of great concern, we concentrate on the six state variables: $\delta_i(t)$, $\omega_i(t)$, $i = 1, 2, 3$. Figures 3.2 plot their angle and speed evolution with time, from $t = 0$ to $t = 0.690$ s, corresponding to a maximum angular deviation of 360° where, presumably, the machines lose synchronism. Note that at $t = 0$, machine angles and speeds are different from zero. Observe also that at $t = 150$ ms, the slope of the speeds experiences a new discontinuity due to the change in the system configuration.

1.5.2   Discussion

The trajectories of the above simulation describing the loss of synchronism between machines are relative to the unperturbed system motion. The question is now how to choose the parameters which mostly influence the system motion. Indeed, a priori the number of parameters is very large, even for such a small system. We could for example consider AVR gains, line impedances, machine electrical characteristics, machine generation powers, and so on.

In what follows, we will choose machine inertias to illustrate the preceding theoretical considerations;[3] nominal values of these inertias are:

$$M_1 = 12.54 \; ; \quad M_2 = 3.39 \; ; \quad M_3 = 1.59 \, .$$

Accordingly, the unperturbed parameter vector associated with the unperturbed motion is

$$\boldsymbol{p} = [M_1, M_2, M_3]^T = [12.54, 3.39, 1.59]^T \, .$$

---

[3] The reason of this choice is that machine inertias provide conclusions easy to interprete.

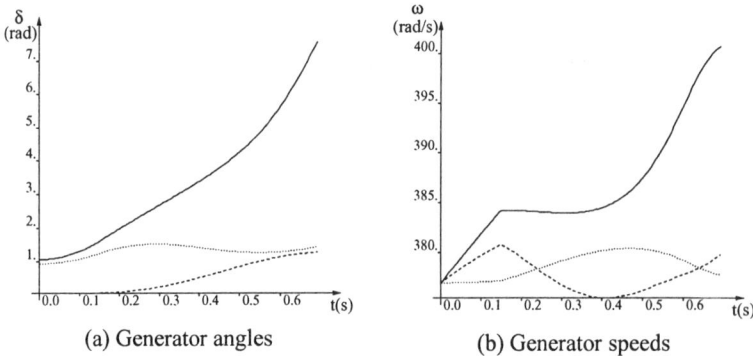

Figure 3.2. Temporal evolution of machine rotor angles and speeds

Note that since the fault application and clearance take place at given time instants ($t = 0$ and 150 ms), the set of equations (3.3) is time-dependent, and should be expressed as:

$$\dot{\mathbf{x}} = \mathbf{F}(\mathbf{x}, \mathbf{p}, t). \tag{3.17}$$

However, one can show that the developments of previous paragraphs dealing with the analysis of the supplementary motion or the synthetic sensitivity function are still valid. In the context of a time-dependent non-linear system, the inertias variations can be referred to as α-variations.[4]

An alternative approach to the analysis of this time-dependent system is to treat the various time periods separately in order to get a succession of time-independent systems described by eq. (3.3). The analysis would consist of first studying the influence of parameter variations in the during-fault period ($t \in]0, 0.150]$) with initial conditions given by the equilibrium point of the initial system configuration. This is a typical α-variation study. Then, the study would continue with the influence of parameter variations in the final configuration. This study is a combination of β-variations (the final state of the during-fault period which depends on the parameter values in the initial state of the final period) and α-variations. A difficulty appears here because the study of β-variations depends on α-variations analysis of the during-fault period.

Although such an analysis of the various time periods separately can be interesting, it will not be pursued here because of the above difficulties.

---

[4]The initial state is $x(0)$. Because we open a line to clear the fault, the new equilibrium point of the system is different from the initial one. So even with a zero fault duration, we will observe changes in the values of the state variables.

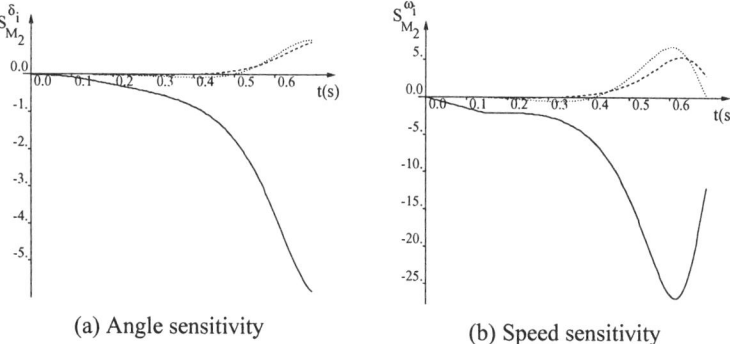

Figure 3.3. Sensitivity analysis of the supplementary motion: influence of machine m_2 inertia on machine rotor angles and speeds: ———— : δ_2, ω_2 ; - - - - : δ_1, ω_1 ; ...... : δ_3, ω_3

1.5.3 Supplementary motion of state variables

The sensitivity study around the initial set point relies on eq. (3.14). The considered state variables are the machine angles $\delta_1, \delta_2, \delta_3$ and speeds $\omega_1, \omega_2, \omega_3$; the considered parameter is the inertia of the most advanced machine m_2 .[5] To compute the sensitivity of the trajectory around the nominal value of M_2 we consider the initial parameter vector

$$p = [12.54,\ 3.39,\ 1.59]^T$$

then perform another simulation under the same conditions but with a small change of the parameter vector

$$\Delta p = [0.,\ 0.05,\ 0.] \ .$$

For example, let us comment on the sensitivity function related to state variable δ_2 (solid line curve of Fig. 3.3a). The curve has been computed using the function[6]

$$S_{M_2}^{\delta_2} = \left.\frac{\Delta \delta_2}{\Delta M_2}\right|_{M_2 = M_{2nom}} = \frac{\tilde{\delta}_2(t) - \delta_2(t)}{0.05} \ .$$

This sensitivity function is zero at the beginning of the simulation (the initial state is the same for the perturbed and the unperturbed motions) and strictly negative otherwise. An increase in M_2 decreases δ_2 . Since the loss of synchronism is caused by machine m_2 going out of step, an increase in M_2 is beneficial to transient stability provided that this has only a second order influence on machines m_1 and m_3 .

---

[5] Actually, according to SIME, m_2 is the CM for the considered contingency.
[6] The sensitivity functions relative to the other state variables $\delta_1, \delta_3, \omega_1, \omega_2, \omega_3$ take on similar expressions.

76 TRANSIENT STABILITY OF POWER SYSTEMS

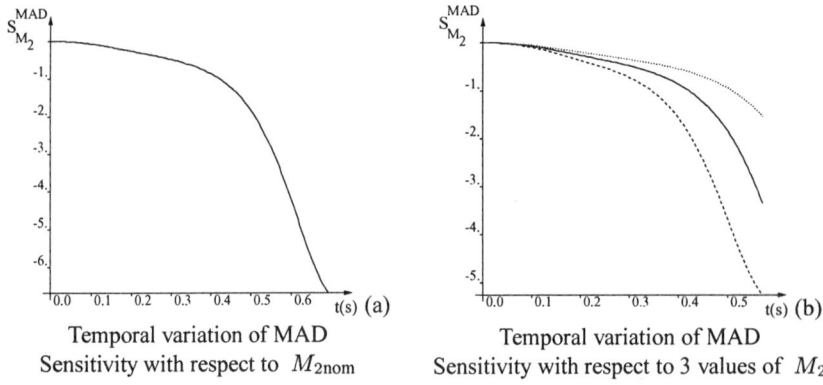

Figure 3.4. Sensitivity analysis of MAD vs M_2

1.5.4 Supplementary motion of time-varying ssf

The variable used here is the maximum angular deviation (MAD in short).[7] The parameter vector related to the unperturbed motion is again composed of the machines' inertia coefficients.

At each time instant, the difference between MAD of the perturbed and the unperturbed motions divided by the variation of M_2 (0.05) provides a time-dependent synthetic sensitivity function, plotted in Fig. 3.4a:

$$S_{M_2}^{MAD} = \left.\frac{\Delta(MAD)}{\Delta M_2}\right|_{M_2=M_{2\text{nom}}}.$$

Inspection of Fig. 3.4a suggests that an increase in M_2 tends to decrease MAD and hence to improve system's stability.

Figure 3.4b illustrates the synthetic sensitivity function dependence on the parameter value around the nominal value:

- the dotted-line curve represents the ssf for $p = [12.54, 3.89, 1.59]^T$
- the solid-line curve represents the ssf for $p = [12.54, 3.39, 1.59]$.
- the dashed-line curve represents the ssf for $p = [12.54, 2.89, 1.59]^T$

This figure shows that the larger the increase in M_2, the larger the decrease in the absolute value of the synthetic sensitivity function. This can be interpreted as follows: a variation of a light machine's inertia has more influence on transient stability than the same variation of a heavier machine.

---

[7]The choice of MAD rather than individual rotor angle/speed state variables is justified by the observation that MAD seems to track well transient stability phenomena.

1.5.5 Discussion

The above two paragraphs performed sensitivity analyses to assess the influence on the 3-machine system stability of the critical machine's inertia coefficient variations. Both studies have reached the same conclusion, although via different ways, namely, an increase in M_2 reinforces the system stability.

Examination of these investigations call for many observations. Below we quote only some salient ones.

Computational burden involved. Both studies rely on two T-D simulations carried out from $t = 0$ (contingency inception) till $t = 690$ ms (system loss of synchronism). The sensitivity analysis involving state variables has in addition required about 150 sensitivity coefficient computations (150 time steps in the time interval $0 - 690$) per state variable and for one value of M_2. The sensitivity analysis involving MAD has required the same number (150) of sensitivity coefficient computations for one value of M_2.

Admittedly, such computations are quite trivial. We only mentioned them to fix ideas.

Sensitivity analysis involving state variables. In the 3-machine system, the machine responsible for the system loss of synchronism is quite easy to identify. In a large-scale system the phenomena can be very unclear. The possible existence of multiswing phenomena are likely to add even more burden: the simple question of exploring the impact of machines' inertias on the system stability may then become inextricable, unless a method like SIME contributes to identify the CMs and also their degree of criticalness.

Sensitivity analysis involving MAD. Here, interpretations become easier. But their validity relies on the validity of MAD as a reliable performance index. Besides, the time-dependence of MAD is another source of confusion. Again, SIME can contribute with its stability margin; indeed, not only is this margin a fully reliable stability index but, in addition, it is time-independent.

Interpretation of parameters' impact on stability. This interpretation often depends on whether a parameter is linked to the CMs or the NMs of the system; or whether the parameter is electrically close to CMs or to NMs. Again SIME can help significantly.

Limitation of sensitivity analyses of the supplementary motion. These studies relying on the supplementary motion deal with small parameter variations. But what about finite (large) parameter variations ? For such investigations SIME provides other types of techniques, discussed in Section 2.

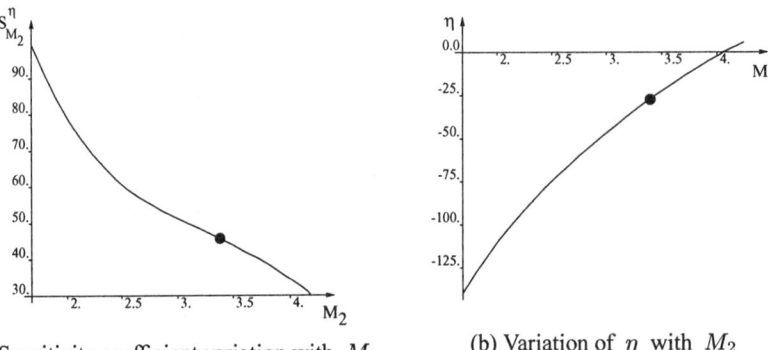

(a) Sensitivity coefficient variation with M_2 (b) Variation of η with M_2

Figure 3.5. Sensitivity analysis of η vs M_2

1.6 Supplementary motion of time-invariant ssfs

The use of SIME's stability margins provides sound synthetic sensitivity functions and makes sensitivity analysis of the supplementary motion a powerful technique in transient stability studies.

Indeed, by its very definition, the margin provides time-independent ssfs and, what is even more important, faithful replicas of the system behaviour.

Denoting η the margin associated with the unperturbed motion and $\tilde{\eta}$ that of the perturbed motion, we consider ssfs expressed by:

$$S_p^\eta = \frac{\tilde{\eta} - \eta}{\Delta p}$$

where p is the parameter responsible for the difference between the two motions.

We illustrate this approach with the same parameter vector as in § 1.5.4. Replacing MAD by η we get

$$S_{M_2}^\eta \approx \left.\frac{\partial \eta}{\partial M_2}\right|_{M_2=M_{2\mathrm{nom}}} = \frac{\tilde{\eta} - \eta}{\Delta M_2} = \left.\frac{\Delta \eta}{\Delta M_2}\right|_{M_2=M_{2\mathrm{nom}}}.$$

Using this procedure, we compute the value of the ssf for

$$\tilde{p} = p + [0,\, 0.05,\, 0]^T = [12.54,\, 3.44,\, 1.59]^T$$

and find 46.74 (> 0) (see in Fig. 3.5a the coordinate of the "bullet"): thus, an increase in M_2 will increase the stability margin and hence the system robustness.

Note that reaching this conclusion has necessitated a single sensitivity computation (instead of 150 for MAD).

The same procedure but for different values of M_2 where

$$p = [12.54,\, M_2,\, 1.59]^T$$

and $\tilde{p} = p + [0,\, 0.05,\, 0]^T = [12.54,\, M_2 + 0.05,\, 1.59]^T$

furnishes the results represented in Fig. 3.5a. The slope of the curve suggests that the lighter the machine and the more the influence of its inertia on the stability margin.

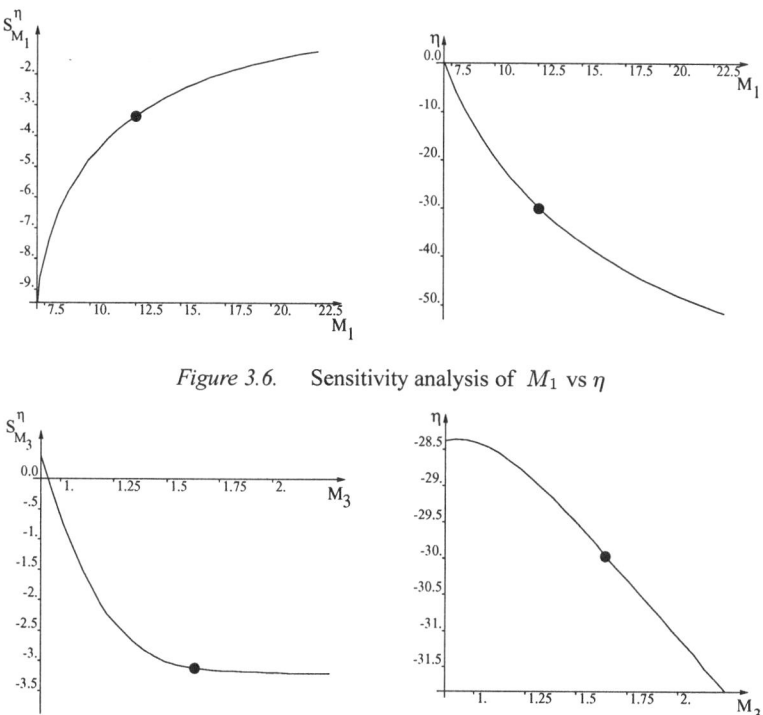

Figure 3.6. Sensitivity analysis of M_1 vs η

Figure 3.7. Sensitivity analysis of M_3 vs η

In Fig. 3.5b, one can observe the margins computed for $M_2 \in [1.68,\, 4.24]$. The slope of the curve at a point gives the corresponding value of the ssf. For values of M_2 larger than 4.06 the system is stable: no loss of synchronism happens anymore. If M_2 is smaller than 1.68, the system is too unstable and SIME is unable to provide a margin.

Now, with respect to the impact of inertias M_1 and M_3 we mention that the value of the above ssf related to a variation of M_1 is -3.33. Since this value is negative, we conclude that an increase in M_1 deteriorates the system stability. The output of the ssf related to M_3 is also negative (-3.01). Hence, variations of inertias of machines m_1 and m_3 have an opposite effect to that of M_2. Figures 3.6 and 3.7 represent curves similar with those drawn for Fig. 3.5 but referring to respectively M_1 and M_3.

In the three-dimensional parameter space ($p = [M_1, M_2, M_3]^T$) the direction that provides the highest rate of increase of the stability region around

$p = [12.54, 3.39, 1.59]^T$ is given by

$$\left[\frac{\partial \eta}{\partial M_1}, \frac{\partial \eta}{\partial M_2}, \frac{\partial \eta}{\partial M_3}\right]^T \equiv [-3.33, 46.74, -3.01]^T .$$

It would thus be possible to get the optimum set of parameters M_1, M_2, M_3 with respect to the contingency considered here. (Note, however, that this would be of academic rather than practical interest.)

2. SIME-BASED SENSITIVITY ANALYSIS

2.1 Specifics and scopes

As pointed out in Chapter 2, the major direct products of SIME are margins and corresponding CMs. These products are the core of all SIME-based sensitivity analyses, whatever their objectives, classified into three types:

- sensitivity analysis per se
- sensitivity analysis for the purpose of stability analysis
- sensitivity analysis for the purpose of control.

By definition, sensitivity analysis per se aims at exploring the influence of parameter variables on system stability. This type of sensitivity analysis may use the method of the supplementary motion in conjunction with synthetic sensitivity coefficients relying on η. Such an exploration was just described in § 1.6.

Sensitivity analyses in the context of stability analysis were already used in Chapter 2; for example, in

- § 3.2.2 to assess the unstable angle δ_u used to calculate a triangle-based stable margin
- § 3.4 to cancel out the $P_{a\,min}$, a substitute for the stability margin η.

More generally, the search of stability limits (critical clearing times (CCTs) or power limits (PLs)) relies on linearized sensitivity analysis, which uses the sensitivity coefficient[8]

$$S_p^\eta \triangleq \frac{\Delta \eta}{\Delta p} = \frac{\eta(k) - \eta(k-1)}{p(k) - p(k-1)} \qquad (3.18)$$

to compute the stability limit

$$p|_{\eta=0} = p(k) - \frac{\eta(k)}{S_p^\eta} . \qquad (3.19)$$

---

[8]Note that, here, margins $\eta(k-1)$ basis and $\eta(k)$ generally correspond to finite (large) variations of parameter p; hence, eqs (3.18), (3.19) hold valid, provided that the margin behaves in a linear fashion.

In the above expressions, parameter p stands for clearing time or OMIB mechanical power, depending upon whether the stability limit sought is CCT or PL.

Finally, sensitivity analysis for the purpose of control addresses the question: how much change to impose on parameter p in order to stabilize the system, i.e., in order to move the system operating conditions from the unstable to the stable region. Again, eqs (3.18), (3.19) form the basis of this exploration. Note, however, that here parameter p may have a broader definition than merely clearing time or OMIB mechanical power.

In some sense, analysis and control pursue sort of complementary objectives, while using a common technique relying on eqs (3.18), (3.19); indeed, analysis aims at determining the distance to the stability region under given operating conditions, whereas control aims at determining the operating conditions under given (generally zero-) distance to the stability region. Stated otherwise, stability analysis deals with the calculation of stability limits relative to a contingency, whereas control deals with means of stabilizing the system with respect to a contingency.

Sensitivity analysis for the purpose of control in general is dealt with in Section 4 of Chapter 4, while next section of this chapter deals with a particular stabilization technique directly inspired by the equal-area criterion.

2.2 On the validity of linearized approximations

We just mentioned that expressions (3.18), (3.19) are the basis for calculating stability limits:

- CCTs for which parameter p represents contingency clearing time, t_e
- PLs for which parameter p represents OMIB mechanical power, P_m.

As will be shown in Chapter 4 these calculations are generally performed using pair-wise linear margins extrapolations (and sometimes interpolations, though quite seldom). The validity of these linear extrapolations relies on the observation that the variation of η with t_e and with P_m is quite linear, provided that

- successive simulations $(k-1)$ and k are not too far away from each other
- the corresponding margins concern the same OMIB, i.e., the same set of CMs.

The second observation is self-explanatory. Let us also remind that ways to circumvent this condition were proposed in § 3.5 of Chapter 2.

With respect to the first observation, it would be hardly possible to give a mathematical proof of the linear behaviour of η vs t_e and η vs P_m, given their extremely intricate relationships. Note, however, the following.

82   TRANSIENT STABILITY OF POWER SYSTEMS

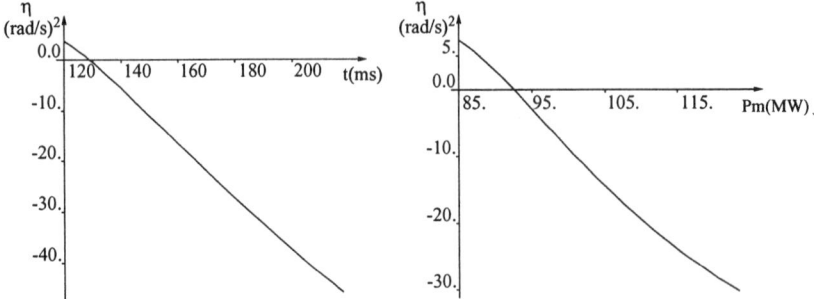

Figure 3.8.   Variation of margin with contingency clearing time and OMIB mechanical power. 3-machine system

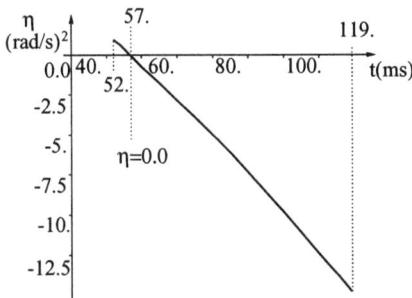

Figure 3.9.   Variation of margin with contingency clearing time. EPRI 627-machine test system

(i) The near linear character of η vs t_e and vs P_m is corroborated by large-scale simulations conducted on a large variety of power systems whose size, in terms of number of machines, varies between 3 and over 600. Figures 3.8 and 3.9 give a sample of typical curves plotted with respectively the 3-machine and the 627-machine EPRI test systems.

(ii) In practice, the interesting range of variation of t_e and P_m is rather restricted. Besides, this range is decomposed so as to conduct iteratively pair-wise extrapolations as described in Section 2 of Chapter 4. This requires only piecewise linear conditions.

A final remark: the non-satisfaction of the above linear approximation used to compute the solution sought (CCT or PL) would yield a larger number of simulations to reach this solution, and hence would affect computational efficiency; but it would not affect accuracy.

3. COMPENSATION SCHEMES (CSs)
3.1 General scope and principle

A given stability scenario is defined by the (pre-fault) operating conditions and the contingency of concern; in turn, this latter may be defined by its type, location and sequence of events including the last clearing time, i.e. the time when the system enters its post-fault configuration. Note that, generally, the severity of a contingency depends strongly on the clearing scenario which may weaken the system's post-fault configuration, as, for example, when tripping many lines connected to the faulted bus.

Stabilizing severe contingencies likely to threaten the power system stability is a task of paramount importance. SIME tackles it by using the large variety of tools that it produces: margins, CMs, and the very OMIB-based equal-area criterion (EAC).

This section focuses on a particular approach, directly inspired by EAC, and summarized in the following two propositions:

- the instability of a multimachine power system is measured by the OMIB margin;
- stabilizing an unstable case consists of canceling out this margin, i.e. of increasing the decelerating area and/or decreasing the accelerating area in the OMIB $P - \delta$ plane (e.g., see Fig. 2.1b).

Broadly, stabilization may be achieved by changing either the contingency scenario (more specifically its clearing time) under preassigned operating conditions, or the operating conditions under preassigned contingency scenario. These changes can be suggested by EAC, by designing "compensation schemes" (CSs) able to appraise approximately the two types of stability limits, namely:

- either critical clearing time (CCT) by adjusting the OMIB P_e curves
- or power limit (PL) by adjusting the OMIB P_m curve.

The following paragraphs deal with the design of these two types of CSs.

3.2 CSs appraising critical clearing times
3.2.1 Description

This paragraph addresses the question: given an unstable case, characterized by the contingency clearing time, t_e and appraised by its negative margin, determine the clearing time decrease necessary to stabilize the case, i.e. to cancel out the margin; in other words, determine the critical clearing time (CCT).

Note that in Chapter 4, CCTs are computed iteratively, in sensitivity-based ways. Here the approximate computation relies on a one-shot EAC-based fashion; it consists of adjusting the accelerating/decelerating areas of the OMIB $P - \delta$ plane by acting on the OMIB electrical power, P_e, so as to compensate for the negative margin.

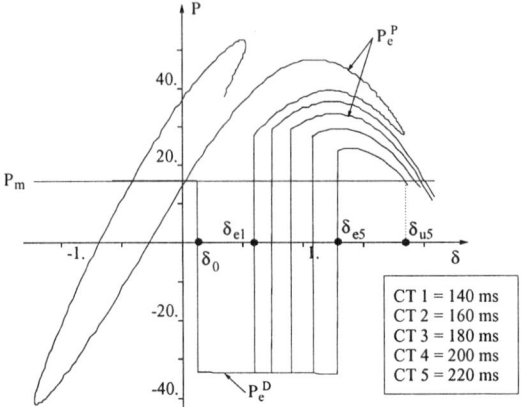

Figure 3.10. Curves P_m, P_e vs $\delta_e(t_e)$ for stability case Nr 672 of the Brazilian system (CCT=141 ms). Adapted from [Ernst et al., 1998a].

Four compensation schemes (CSs) are tentatively proposed. They are suggested by the shape of the P_e curves variation with t_e (or equivalently with the clearing angle δ_e); such typical variations are displayed in Fig. 3.10 for various clearing conditions. These curves were plotted for a contingency simulated on the Brazilian system described in Section 4 of Appendix B, on which the proposed CSs were initially tested.

Figures 3.11 display these CSs[9], along with the margin, represented by the shaded rectangular area; it is computed for $t_e = 167$ ms and is equal to -14.7 (rad/s)^2. The margin is compensated in four different ways of increasing the decelerating and/or decreasing the accelerating areas. They yield four different approximate critical clearing angles, δ_{ci}'s, and hence approximate critical clearing times, t_{ci}'s. Note that Figs 3.11a to 3.11d yield increasing t_{ci} values: 125, 136, 143 and 144 ms respectively (vs 141 ms, which is the actual CCT).

Generally, it is quite likely that the closer the clearing time to the actual CCT and the better the approximate t_{ci}'s. Also, the degree of modeling sophistication of a power system is likely to influence the performances of the CSs.

CSs may be useful in various practical applications, in particular, the following two.

---

[9]The compensation schemes in Figs 3.11b and c were proposed by [Zhang, 1996]

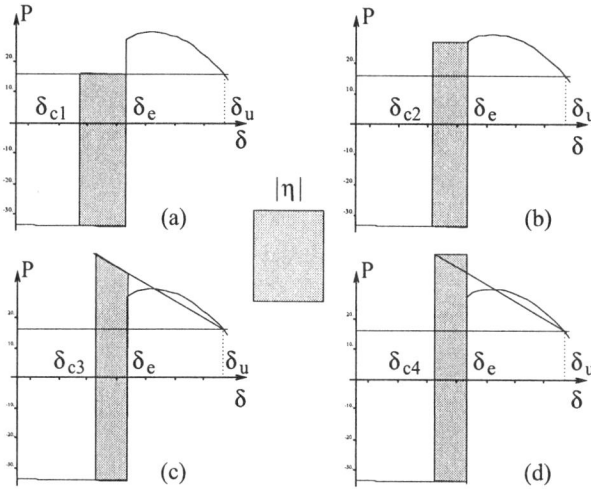

Figure 3.11. Four compensation schemes for CCT assessment. Stability case Nr 672. $t_e = 167$ ms ($\delta_e = 0.82$ rad); $\eta = -14.7$. Taken from [Ernst et al., 1998b]

- For filtering purposes, whenever the clearing time, t_e, is not too far away from the actual CCT, the CSs may save one out of the two simulations used by the two-margin extrapolation filter of Chapter 4, § 2.4.2. The resulting CPU gains are particularly interesting for simulations run with detailed power system models. [10]
- For CCT assessment, whenever the initial t_e is far away from (i.e. much larger than) the actual CCT, using new t_e's suggested by the CSs may contribute to speed up significantly the convergence procedure with respect to the pure extrapolation procedure described in § 2.2 of Chapter 4.

A final remark: the CPU times needed for CSs computations per se are virtually negligible.

3.2.2 Illustrative examples

In this paragraph, the CSs described in Figs 3.11 are tested on two sets of stability cases relative to two power systems.

The one is the 3-machine system, described in Section 1 of Appendix B. Twelve contingencies are simulated and their approximate t_{ci}'s are computed by using the CSs, with a clearing time of 200 ms; these values are compared with the accurate values computed by SIME in Chapter 4, § 2.4, Table 4.8.

Table 3.1 collects the results. Note that for this clearing time, contingency Nr 2 has no margin, while the last 6 contingencies are shown to be stable. As can be seen in the table, t_{c3} and t_{c4} are closer to the actual CCT of Column 3. But the number of cases is not sufficient to draw conclusions.

---

[10] The gain is even more important when the OMIB structure changes from one simulation to the other(s).

86 TRANSIENT STABILITY OF POWER SYSTEMS

Table 3.1. CCTs of the 3-machine system. $t_e = 200$ ms

1	2				3
Cont. Nr	Compensation schemes (ms)				SIME (ms)
	t_{c1}	t_{c2}	t_{c3}	t_{c4}	
2	/	/	/	/	95
3	36	69	88	98	130
7	49	65	94	101	147
6	162	180	181	181	189
1	186	186	194	194	194
4	194	193	196	196	196
5	S	S	S	S	220
10	S	S	S	S	225
11	S	S	S	S	227
8	S	S	S	S	251
12	S	S	S	S	272
9	S	S	S	S	297

Table 3.2. CCTs of the Brazilian system. $t_e = 200$ ms

1	2	3				4	5
Cont. Nr	η	Compensation schemes				SIME (ms)	T-D (ms)
		t_{c1}	t_{c2}	t_{c3}	t_{c4}		
687	−47.3	50	−	126	134	119	117
656	−36.8	57	12	130	137	133	132
686	−17.9	136	149	174	175	136	138
672	−39.4	82	116	132	137	141	143
630	−30.9	95	105	153	157	146	147
680	−35.6	100	132	145	149	149	152
708	−34.9	103	134	148	151	149	152
598	−35.0	103	134	148	152	150	152
629	−30.8	95	107	153	157	153	152
463	−34.6	104	136	149	152	154	152
692	−29.1	103	116	157	160	156	157
707	−33.6	108	139	151	154	157	157
690	−22.2	117	126	164	166	161	162
437	−69.3	110	151	159	163	165	162
434	−64.0	119	158	165	168	170	168
689	−16.4	143	155	177	178	173	173
628	−16.8	141	153	176	177	173	173
522	−57.3	130	166	171	173	175	173
362	−15.8	145	157	178	179	177	178
516	−43.3	150	177	180	181	183	183

Chapter 3 - SENSITIVITY ANALYSIS 87

The second set of simulations is performed on the Brazilian system described in Section 4 of Appendix B. The results collected in Table 3.2 report on the simulation of 20 contingencies; their CCTs are computed in three different ways:

- by means of the 4 CSs for a clearing time of 200 ms; see Columns 3; the corresponding margin is listed in Column 2
- by using the algorithm of § 2.2 of Chapter 4; the results are listed in Column 4, labeled "SIME"
- by using the T-D program ST-600 [Valette et al., 1987]; see Column 4 labeled "T-D"

The table shows that:

- t_{c1} provides a lower bound of the actual CCT (though sometimes much smaller than it);
- t_{c2} behaves in a similar way, though with some failures;
- t_{c3} is sometimes larger and sometimes smaller than CCT;
- t_{c4} is generally larger than the actual CCT; and whenever smaller, it is very close to it. It may thus provide an upper CCT bound.

The above observations suggest in particular that:

- t_{c1} might be used as a "safeguard", i.e. to avoid missing contingencies in the context of contingency filtering;;
- t_{c2}, t_{c3} show to be less interesting;
- t_{c4} seems quite suitable, in particular for screening contingencies.

Finally, observe the good agreement of SIME with the T-D program of Hydro-Québec.

3.3 CS appraising power limits

3.3.1 Principle

This paragraph addresses the question: given an unstable case characterized by its negative margin, determine the decrease in generation power necessary to stabilize it, i.e., to cancel out this margin. This question will be subdivided in the following two tasks: (i) devise a CS which assesses the necessary variation of the OMIB mechanical power; (ii) report this power on system machine generation.

3.3.2 Power limit of OMIB

Figure 3.12 helps answer above question (i). It shows that lowering the mechanical power from P_m to P'_m modifies the accelerating and decelerating areas of OMIB $P - \delta$ representation. More precisely, it increases the decelerating area (by areas A_2 to A_5), decreases the accelerating area (A_1 and A_6), and increases the accelerating area (A_7). Thus,

$$|\eta| = \sum_{i=1}^{6} A_i - A_7 . \qquad (3.20)$$

Expression (3.20) links directly $|\eta|$ to the amount of P_m decrease. Note that, actually, decreasing P_m implies increasing P_e^P, thus adding one more

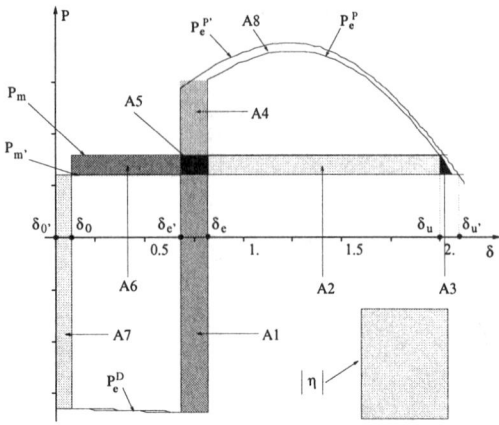

Figure 3.12. Compensation scheme for controlling generation. Stability case Nr 672 of the Brazilian system. $t_e = 167$ ms ; $\eta = -14.7$; $P_m = 324$ MW ; $P_C = 1050$ MW. Taken from [Ernst et al., 1998b]

decelerating area, say A_8, comprised between curves $P_e^{P'}$ and P_e^{P}. But its computation would imply another T-D simulation. On the other hand, neglecting A_8 gives a pessimistic assessment, i.e. an overestimation of $\Delta P_m = P_m - P_{m'}$, which, however, is rather minor (a small percentage).

Computation of areas $(A_1$ to $A_7)$ is elaborated below, in § 3.3.5.

3.3.3 Power limits of system machines

With regard to above question (ii), observe that the distribution of the OMIB ΔP_m among actual machines' generation must obey eq. (2.9). This equation suggests that decreasing P_m implies decreasing generation in critical machines (CMs) and increasing generation in non-critical machines (NMs) according to:

$$-\Delta P_m = -\frac{M_N}{M_N + M_C}\Delta P_C + \frac{M_C}{M_N + M_C}\Delta P_N \qquad (3.21)$$

where

$$\Delta P_C = \sum_{k \in C} P_{mk} \; ; \; \Delta P_N = \sum_{j \in N} P_{mj} . \qquad (3.22)$$

Further, when the concern is to keep unchanged the overall system generation, the condition[11]

$$\Delta P_N = -\Delta P_C \qquad (3.23)$$

---

[11] neglecting variation of losses.

yields, according to eq. (3.21):

$$\Delta P_N = -\Delta P_C = -\Delta P_m = \Delta P. \qquad (3.24)$$

In the example case of contingency Nr 672 considered in Fig. 3.12, the computation of ΔP_m via expression (3.20) yields 91 MW, to be compared with 85 MW provided by a pure (but much more tedious) T-D procedure[12]. Hence, according to eq. (3.24), the generation of the critical machine (Nr 2712) will be decreased by 91 MW and that of non-critical machines increased by the same amount.

3.3.4 Discussion

Equations (3.20), (3.24) govern any stabilization procedure relying on active power *reallocation (shift)*. Their inspection calls for many observations.

1.- The value ΔP_m computed via (3.20) (and hence ΔP_C) is only approximate; the accurate value (corresponding to $\eta = 0$), will be computed via the iterative procedure described in Chapter 4.

2.- A priori, eq. (3.24) may receive a (very) large number of solutions, depending on the number of CMs and NMs.

3.- It will therefore be possible to derive various solutions able to encounter various requirements.

4.- More specifically, transient stability requirements will generally impose conditions on CMs, while other types of requirements (e.g., market's requirements) can be met via NMs' rescheduling. These issues are addressed in Chapter 4.

3.3.5 Numerical example

Let us use the compensation scheme for controlling the generation power in the 3-machine system, subjected successively to contingencies Nr 2 and 3.

Table 3.3 summarizes the results obtained for a clearing time of 150 ms.

Column 3 indicates the corresponding margin, while Column 5 indicates the power decrease suggested by the CS for stabilizing the case (cancelling out the margin). Columns 4, 6 and 7 list successively the initial power generated by the CM(s), the power corrected according to the CS, and the actual power limit for which the margin is indeed zero. Comparison of the results of the last two columns shows that the correction provided by the CS is quite good.

Note that the above stabilization may be refined via the iterative procedure of Chapter 4, where more examples will also be reported.

---

[12]The 7% discrepancy is because of neglecting area A_8 between curves $P_e^{P'}$ and P_e^{P}; indeed, the computation of this area yields 6 MW.

90 TRANSIENT STABILITY OF POWER SYSTEMS

Table 3.3. Contingency stabilization via the CS of Fig. 3.12. $t_e = 150$ ms

Cont. Nr	CMs	Margin	P_C (MW) (initial)	ΔP_C (MW) (CS)	P_C (MW) (CS)	P_{CL} (MW) (actual)
1	2	3	4	5	6	7
2	m_2, m_3	−17.55	248	−33	215	219
3	m_2	−9.78	163	−15	148	153

3.3.6 Computing areas of the CS

Let us compute areas A_1 to A_7, under the assumption that the P_e^P curve variation with t_e is negligible and that the mechanical power P_m remains constant during transients.

According to Fig. (3.12), A_1 and A_7 are expressed respectively by

$$A_1 = (\delta_e - \delta'_e)(P_m - \Delta P) \tag{3.25}$$
$$A_7 = (\delta_o - \delta'_o)(P_m - \Delta P) \tag{3.26}$$

where δ_e, δ'_e may be derived from eq. (2.2) as:

$$\delta_e = (P_m - P_{eD})\frac{t_e^2}{2M} + \delta_o \tag{3.27}$$

$$\delta'_e = (P_m - \Delta P - P_{eD})\frac{t_e^2}{2M} + \delta'_o . \tag{3.28}$$

Therefore, eqs (3.25), (3.26) yield

$$A_1 - A_7 = (P_m - \Delta P)\frac{t_e^2}{2M}\Delta P . \tag{3.29}$$

On the other hand, inspection of Fig. 3.12 shows that

$$A_2 + A_5 + A_6 = \Delta P (\delta_u - \delta_o) \tag{3.30}$$

and that

$$A_3 = \frac{1}{2}c(\Delta P)^2 \tag{3.31}$$

where c is the slope of the curve P_e^P at δ_u.

Finally, Fig. 3.12 and eqs (3.27), (3.28) yield:

$$A_4 = (\delta_e - \delta'_e)(P_e^P - P_m) = \left(\delta_o - \delta'_o + \Delta P\frac{t_e^2}{2}\right)(P_e^P - P_m) . \tag{3.32}$$

In this latter expression, δ'_o is not known. Hence, A_4 may either be approximated by neglecting the corresponding term and writing

$$A_4 \simeq \Delta P \frac{t_e^2}{2}(P_e^P - P_m) \qquad (3.33)$$

or expressed correctly by computing δ'_o. In turn, this latter computation requires either a mere sensitivity computation or running an additional load flow and using the following iterative procedure:

(i) compute a first-approximation $\Delta P^{(1)}$ by solving eq. (3.20) with A_4 expressed by (3.33);
(ii) decrease the generation power in the critical machines by $\Delta P^{(1)}$ and increase accordingly in non-critical machines; run a new LF with these new power levels; let $\delta'^{(1)}_o$ be the corresponding prefault OMIB angle;
(iii) compute a new value for A_4, using (3.32) with $\delta'_o = \delta'^{(1)}_o$;
(iv) solve eq. (3.20) for ΔP; set

$$\Delta P^{(2)} = \Delta P \, ;$$

if $\Delta P^{(2)}$ is close enough to $\Delta P^{(1)}$, stop;
otherwise, set

$$\Delta P^{(1)} = \Delta P^{(2)}$$

and go to (ii).

4. SUMMARY

Elements of conventional sensitivity analysis methods have first been recalled. The analysis of the supplementary motion was found to suit transient stability concerns, provided an appropriate performance index can be defined to derive time-invariant "synthetic sensitivity functions". Such an index is the stability margin provided by SIME.

Three application contexts of sensitivity analysis were then identified, namely: (i) sensitivity analysis per se; (ii) sensitivity analysis for the purpose of stability analysis; (iii) sensitivity analysis for the purpose of control.

Type (i) can be performed using the supplementary motion method.

Types (ii) and (iii) rely on first order sensitivity coefficients appraising margin vs finite parameter variations. Such parameters of concern are contingency clearing time and OMIB mechanical power. Pair-wise extrapolations of linear margins will be used in Chapter 4 to compute contingency critical clearing times (CCTs) and power limits (PLs). These compensation schemes will further be used in Chapter 4, in particular to devise control techniques.

Finally, two compensation schemes were proposed, yielding one-shot procedures for computing approximate CCTs and PLs. Such schemes will further be used in Chapter 4, in particular to devise control techniques.

Chapter 4

PREVENTIVE ANALYSIS AND CONTROL

Chapter 2 has dealt with calculation of margins and identification of critical machines. Chapter 3 has used them to derive sensitivity analysis techniques. This chapter shows that stability margins, critical machines and sensitivity analysis are the necessary and sufficient ingredients for conducting any type of transient stability study, including analysis, filtering, ranking, assessment and stabilization of contingencies. More precisely, the objectives of Chapter 4 are:

- *to derive a systematic procedure for computing stability limits (critical clearing times and power limits);*
- *to describe a variety of techniques able to screen contingencies and select the "interesting", i.e., the constraining ones;*
- *to set up a unified approach to contingency filtering, ranking and assessment;*
- *to derive systematic procedures for stabilizing the constraining contingencies;*
- *to pave the way towards on-line TSA&C techniques, able to meet various needs already existing or emerging in the restructuring electricity market. These techniques will be elaborated in Chapter 5.*

1. PRELIMINARIES

1.1 Chapter overview

This chapter covers all aspects of transient stability assessment (TSA) and control in the preventive mode.

TSA aims at determining whether the system is able to withstand plausible contingencies and, if not, at measuring the severity of those found to be "harmful", i.e. able to cause system's loss of synchronism, should they occur. Hence, TSA addresses the twofold problem: contingency filtering and analysis.

In conventional transient stability studies, analysis aims at measuring contingency severity in terms of stability limits (SLs). The SLs of current concern are critical clearing times (CCTs) and power limits (PLs). Recall that for a contingency applied under preassigned operating conditions, CCT is the maximum time that the contingency may remain (i.e. the maximum clearing time), without the system losing its capability to recover a synchronous operation. The precise definition of a PL is more difficult. Loosely, we will say that for a given contingency applied under preassigned duration (clearing time), PL is the maximum power that the system may sustain without losing synchronism. From a physical point of view, it is clear that PL depends on a large number of factors, and in particular on generation patterns of the system power plants. Hence, it is not a univocally defined quantity: different procedures will yield different PLs, as shown in § 2.3.

The choice between these two "measures" of the system robustness vis-à-vis a contingency is a matter of operating practices, which in turn are linked to system specifics and physical limitations. Generally, CCTs are more popular in Europe, while PLs are preferred in the United States. As will be shown below, they are both assessed by SIME on the basis of margins and sensitivity techniques developed in Section 2. But while following the same pattern, CCTs will appear to be more straightforward to assess than PLs.

Section 2 of this chapter is devoted to the computation of SLs.

Contingency filtering, on the other hand, is the indispensable complement of stability analysis. Indeed, in practice, accurate assessment of a stability limit (be it CCT or PL) is interesting only for "severe" or "constraining" contingencies.[1] Hence, prior to analysis, it appears convenient to discard all "mild" contingencies on the basis of a faster, though less accurate procedure. This is precisely the role of contingency filtering techniques. And since, as just mentioned, CCTs are easier to compute than PLs, contingency filtering techniques will generally rely on approximate assessment of CCTs. Stability margins will also be the basis of this assessment.

In practice, it is interesting to rank the contingencies selected by the filtering phase according to their severity and, further, to assess in a finer way the severest among them. Again, contingency assessment uses stability margins.

Section 3 proposes an integrated technique called FILTRA (for FILTering, Ranking and Assessment of contingencies). FILTRA is made up of two blocks adaptable to power system specific features and needs.

Another paramount achievement of SIME is control. Its objective is to "stabilize" harmful contingencies (processed one-by-one or simultaneously), i.e., to design proper countermeasures which modify the system operating

---

[1] The different classes of contingencies are defined in § 3.2.

conditions so as to render the contingencies harmless. Stabilization procedures are developed in Section 4, on the basis of margins and CMs.

A final, general reminder: without exception, use of the Preventive SIME and of all derived techniques implies its coupling and combined use with a transient stability T-D program.

1.2 A measure for assessing computing performances

This chapter is the first to deal with SIME's practical applications to real-world problems. For such applications, computing performances is an issue of great concern; it becomes crucial when it comes to real-time operation.

We therefore need a "unit" able to appraise the computing performances of the various SIME based techniques. And, since SIME is coupled with a T-D program, it is interesting to define a "unit" in terms of the computing effort required by the corresponding T-D program.

This "unit" relies on the observation that the CPU time imputable to SIME itself is virtually negligible vis-à-vis any other task involved. (To fix ideas, it corresponds to about one iteration of the power flow program.) Hence, the overall computing effort reduces to that for running the T-D program during the short periods required by SIME. Therefore, a handy means for comparisons appears to be the corresponding seconds of Time-Domain Integration (sTDI for short).

This "measure" renders comparisons independent of the computer used and of the system size. [2]

2. STABILITY LIMITS
2.1 Basic concepts

The computation of stability limits (CCTs or PLs) follows the general pattern of SIME-based sensitivity analysis expressed by eqs (3.18), (3.19) of Chapter 3.

1. According to SIME, a stability limit (SL) corresponds to zero margin conditions.

2. Since SIME relies on the OMIB concept, i.e. on unstable multimachine trajectories[3], the search of zero-margin conditions will consist of using successive unstable simulations of decreasing severity.

3. Further, sensitivity considerations suggest to process pair-wise margins' extra-(inter-)polations.

4. Recall also that unstable margins exist only for a limited range of conditions (see § 3.3 of Chapter 2). For very unstable simulations, § 3.4 of Chapter 2

---

[2] Note, however, that for T-D programs with variable stepsize, the number of sTDIs may depend on the simulation range.
[3] Recall that, strictly speaking, an OMIB is defined on unstable scenarios only.

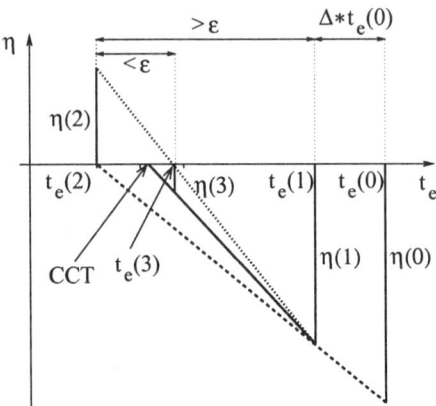

Figure 4.1.   Schematic description of the CCT search procedure

has proposed sensitivity calculations yielding conditions within the range of existing margins; nevertheless, to reduce the number of simulations as much as possible, usage will be made of "hints" able to judiciously choose initial conditions.[4]

5. Finally, recall that to detect multiswing instabilities (whenever they are of interest) a first-swing stable simulation should tentatively be carried out for the entire integration period: either the simulation will be stable throughout, showing that the system is also multiswing stable; or it will meet multiswing instabilities. In this latter case, further simulations should be carried out under milder conditions, until reaching fully stable ones. (Recall that multiswing phenomena are addressed in Section 5 of Chapter 2.)

2.2   Critical clearing times

2.2.1   Basic procedure

For a given contingency and pre-fault operating conditions, the above considerations lead to the iterative computation summarized below and described in Fig. 4.1. The rationale and choice of parameter Δ is discussed in § 2.2.2.

---

[4] Ideally, these conditions should drive the system in the range of margins' existence (see § 3.3 of Chapter 2). In other words, they should be severe enough to yield negative margins, yet not too severe to avoid going beyond the range of margins' existence. This issue is considered below, in § 2.2.2.

(i) Set $k = 0$ and choose $t_e(k)$ so as to get an unstable simulation.[5] Compute the corresponding margin $\eta(k) < 0$. Set $k = k+1$ and $t_e(k) = t_e(k-1) - \Delta * t_e(k-1)$.

(ii) Run a T-D simulation with $t_e(k)$ as clearing time. If $\eta(k) > 0$, set $k_{st} = k$. If $\eta(k) < 0$, set $k_u = k$.

(iii)

- if $\eta(k) < 0$ extrapolate the margins of the two last unstable simulations to get $CCT(k)$.
- if $\eta(k) > 0$ interpolate between the last stable and the last unstable simulations to get $CCT(k)$.

(iv)

- If no stable simulation exists, set $k = k+1$, $t_e(k) = CCT(k-1)$ and go to (ii).
- If a stable simulation exists, and if $t_e(k_u) - t_e(k_{st}) > \epsilon$, set $k = k+1$, $t_e(k) = CCT(k-1)$ and go to (ii); otherwise go to (v).

(v) The value of the critical clearing time is $CCT(k)$.

In short, the overall CCT search consists of using pair-wise linear margins' extra-(inter-)polations iteratively, until getting the accuracy sought.

Figure 4.1 gives a schematic description of the basic algorithm. Many other variants and extensions may be thought of, as, for instance, those used in the illustrative examples of §§ 2.2.4, 2.2.5.

2.2.2 Parameters and technicalities

1.- Parameter Δ is chosen small enough to preserve the validity of margins' linear inter-(extra-)polation, yet large enough to restrict the number of simulations. Note that Δ may take on different values for different ranges of t_e's. For example, $\Delta = 10\%$ for t_e ranging in between 50 and 200 ms, and $\Delta = 20\%$ for $t_e > 200$ ms.

2.- Parameter ϵ is taken quite small (e.g. $\epsilon \approx 2.5\%$). Indeed, we want the stable simulation to be close enough to the last unstable one so as to use, by continuation, the same OMIB.

3.- In essence, the above margin-based iterative procedure is much more effective than the "blind" search of a pure T-D simulation, be it dichotomic or not.

---

[5]The cases where $t_e(0)$ does not yield an unstable (negative) margin are treated below, in § 2.2.3.

4.- Finally, let us again stress that the algorithm of § 2.2.1 describes just method's very principle. Many refinements can be thought of, and are actually used.

2.2.3 Initial clearing time conditions

It has been emphasized that the proper choice of the initial clearing time ($t_e\,(k=0)$) contributes to the efficiency of the overall procedure[6]. Indeed, a too small $t_e(0)$ could yield a stable simulation that SIME cannot treat at once for want of information about the mode of machines' separation; in such a case a new simulation is run using $t_e(1)$ larger than $t_e(0)$ in order to get an unstable margin.[7] On the other hand, a too large $t_e(0)$ would yield an unduly unstable simulation, beyond the range of margins' existence; in such a case, additional sensitivity calculations are needed to recover the "interesting" range of t_e's (see § 3.3 of Chapter 2).

To avoid such inconveniences, the following pragmatic criteria are used to choose $t_e(0)$ automatically. These criteria stop the during-fault T-D simulation and switch to the post-fault one[8] as soon as:

- the maximum angular deviation between extreme system machines reaches a given, preassigned value, $\delta_{ij\,\max}$; or
- the angle of a candidate OMIB reaches a maximum, preassigned value, $\delta_{OMIB\,\max}$; or
- a preassigned maximum clearing time, $t_{e\,\max}$, is reached before any one of the above criteria.

Note that the choice of $\delta_{ij\,\max}$ and $\delta_{OMIB\,\max}$ needs an off-line tuning, which, however, is quite easy. Indeed, experience shows that the range of variation of "optimal" values is quite restricted and rather independent of the system size (whether having 3 or 600 machines).[9] On the other hand, the choice of parameter $t_{e\,\max}$ depends on the particular application sought.

2.2.4 Performances

Computing requirements. The number of iterations needed for the accurate computation of a CCT depends on various parameters: the power system itself[10], the severity of the contingency, the good choice of $t_e(0)$, the size of

---

[6]Note, however, that this choice does not affect accuracy.
[7]Subsequently, the simulation corresponding to $t_e(0)$ is re-run (unless already stored) until reaching t_r for which the corresponding stable margin is computed.
[8]the procedure has to be adjusted in the case there is a sequence of contingency scenarios.
[9]For large-scale systems it is advisable to consider machines rotor angles *relative* to their initial (pre-fault) values.
[10]though not its size.

Chapter 4 - PREVENTIVE ANALYSIS AND CONTROL 99

Δ and, last but not least, the absence or not of multiswing instabilities and interest in their search. A rough estimation based on thousands of simulations (on a large number of power systems and a very large number of contingencies) suggests the following mean values per contingency:

(i) computation of first-swing CCT only: 3.5 simulations, of which 1 may or may not be stable; the stable simulation, if any, is conducted up to t_r (determined by condition (2.16))

(ii) search for existence and computation of multiswing CCT, otherwise of first-swing CCT:

- in the presence of multiswing instabilities, 4.5 unstable and 1 stable simulations performed on the entire integration period
- in the absence of multiswing instabilities, 3.5 unstable and 1 stable simulations performed on the entire integration period.

In terms of sTDI defined in § 1.2, the computing performances of the above different cases rely on the following considerations:

- the computing time needed to run an unstable simulation corresponds to t_u sTDI
- the computing time needed to run a stable simulation corresponds to t_r sTDI or to MIP,[11] depending upon whether the stable simulation is stopped at t_r or is performed on the entire integration period.

Of course, the values of these parameters vary with the very power system, the contingency under consideration and for a given contingency with the severity of the simulation conditions (i.e., with t_e): the larger the t_e, the severer the simulation conditions and the smaller the t_u. As for MIP, it mainly depends on the degree of power system modeling sophistication.

These considerations will be illustrated below, on the example of § 2.2.5.

Accuracy. The accuracy of the above procedure depends on the accuracy of margin computations.[12] And since according to Section 3 of Chapter 2 negative margins are much more accurate - and in fact very accurate - than positive margins, it is advisable to rely on negative margins only: in terms of CPU this is quite unexpensive, since it "costs" an additional unstable simulation, which is significantly less "expensive" than the stable simulation that anyhow is needed to detect possible multiswing phenomena.

---

[11] MIP stands for "Maximum Integration Period". Recall also that t_u corresponds to conditions (2.15) and t_r to conditions (2.16).

[12] Note that the linearized margins inter-(extra-)polation affects the convergence but not the accuracy of the final result. Chapter 3 (see Figs 3.8, 3.9) suggests that this linearized approximation is generally valid.

100   TRANSIENT STABILITY OF POWER SYSTEMS

2.2.5   Illustrations on the 3-machine system

We consider the 3-machine system described in Section 1 of Appendix B and compute the CCT of contingency Nr 3. The search relies on the iterative procedure of § 2.2.1 together with considerations of §§ 2.2.2, 2.2.3. Accordingly, we use the following parameter values:

$\epsilon = 15$ ms ; $\Delta = 10\%$; $\delta_{ij\,max} = 180°$; $\delta_{OMIB\,max} = 110°$; $t_{e\,max} = 500$ ms .

Table 4.1.   Iterative CCT computation. 3-machine system. Contingency Nr 3

1	2	3	4	5
Iter. Nr	Clear. time t_e (ms)	Margin η (rad/s)^2	Extra-(inter)-pol. (ms)	t_u (MIP) (s)
1	218	−44.25	/	0.239
2	198	−34.99	123	0.259
3	123	5.85	134	(3)
4	134	−1.64	**131**	0.510

Table 4.1 summarizes the procedure. It shows that the initial clearing time $t_e(0) = 218$ ms. (Incidentally, here, this time is imposed by the criterion on $\delta_{OMIB\,max}$ which was first met.)

Thus, the system enters its post-fault configuration at $t_e(0) = 218$ ms ; it reaches the instability conditions (2.16) at $t_u = 239$ ms. At this time, the OMIB is found to be composed of 1 CM (m_2) and 2 NMs (m_3 and m_1)[13]; the corresponding (normalized) margin is −44.25 (rad/s)^2.

Table 4.1 displays in bold the result of the iterative procedure: CCT = 131 ms.

On the other hand, the last column of the table lists the successive values of t_u (apart from the stable simulation for which one reads MIP= 3 s .) It shows that t_u varies inversely to t_e, i.e., to the contingency severity. Further, the sum of t_u's and MIP listed in this column provides the total computing time: $3 + 1.008 \simeq 4.01$ sTDI: obviously, a large proportion of this time is devoted to run the stable simulation on the MIP (here 3 s); this is the price to pay for tracking multiswing instabilities. We mention that, here, $t_r = 501$ ms; this suggests that if we are not interested in multiswing phenomena, we can "save" 2.5 sTDI: the overall computing effort then reduces from 4.01 to 1.50 sTDI, which corresponds to speeding up the procedure by a factor of about 2.67.

Figure 4.2 displays the phase plane representation summarizing the above simulations.

In order to compare SIME's accuracy with respect to the T-D program coupled with SIME, we have computed the CCTs of the 12 contingencies with the T-D program run alone and with SIME.

Table 4.2 gathers these results (Columns 2, 3). Note that the T-D program proceeds by dichotomy; the error is bounded in the interval ±2 ms. Obviously, the CCTs provided by T-D and by SIME are virtually the same.

Observing that the dichotomic search requires a quite larger number of (lenghty) stable T-D simulations suggests that SIME allows significant computing savings; this is not one of SIME's major accomplishments, though.

Finally, Column 4 informs about the CMs corresponding to the different contingencies.

---

[13] Incidentally, recall that Fig. 2.1 drawn for contingency Nr 2 has 2 CMs (m_2 and m_3) and 1 NM (m_1).

Chapter 4 - PREVENTIVE ANALYSIS AND CONTROL

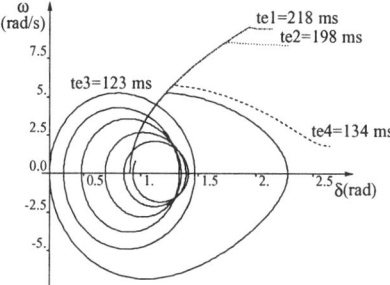

Figure 4.2. Phase plane representation of simulations of Table 4.1. 3-machine system.

Table 4.2. CCTs (in ms) and CMs of the 12 contingencies on the 3-machine system

	1	2	3	4
Contingency identif.	T-D CCT	SIME CCT	Crit. machines	
2	95	95.7	$m_2\ m_3$
3	131	131	m_2
7	148	147	$m_2\ m_3$
6	189	189	m_3
1	192	194	$m_2\ m_3$
4	195	196	m_2
5	218	220	m_2
10	225	225	$m_2\ m_3$
11	227	227	$m_2\ m_3$
8	250	251	$m_2\ m_3$
12	274	272	$m_2\ m_3$
9	297	297	$m_2\ m_3$

2.2.6 Illustrations on the 627-machine system

Let us now consider the EPRI 627-machine test system A. Its data are described in Section 3 of Appendix B. (See also Table 4.9.)

Table 4.3 summarizes the simulation results of CCT computation for contingency Nr 9. Observe that, according to Column 6 of the table, this contingency creates multiswing and backswing instabilities (denoted (B)). Note also that the number of swings changes during the simulations. Finally, notice that, here the times to instability, t_u's are much larger for these multiswing phenomena than for first-swing ones.

The resulting CCT is indicated in bold. We mention that the ETMSP transient stability program run alone gives the same value.

102   TRANSIENT STABILITY OF POWER SYSTEMS

Table 4.3.   Iterative CCT computation. 627-machine system. Contingency Nr 9

1	2	3	4	5	6
Iter. Nr	Clear. time t_e (ms)	Margin η (rad/s)^2	Extra-(inter-)pol. (ms)	t_u (MIP) (s)	Nr of swings
1	180	−0.57	/	3.19	2 (B)
2	171	−2.09	/	7.75	5 (B)
3	162	6.72	**168**	(10.0)	/

Figures 4.3 plot two sets of curves corresponding to two different clearing times: one ($t_e = 171$ ms) yields an unstable simulation, the other ($t_e = 162$ ms) a stable one. Each set consists of respectively the multimachine swing curves,[14] the OMIB swing curve and the OMIB $P - \delta$ representation. From the multimachine swing curves one can see that there is one CM only. The OMIB swing curve, on the other hand, illustrates the backswing and multiswing instability phenomena.

Finally, Fig. 4.4 plots the corresponding phase plane representation of the above simulations. Notice that the somehow additional complexity of this figure with respect to Fig. 4.2 is merely due to multiswing phenomena, absent in the 3-machine system. Other differences between these two figures come from the existence of backswings on the 627-machine system, and of the modeling details.

2.3   Power limits
2.3.1   Preliminaries

Basically, the general pattern of § 2.2 dealing with CCTs applies also to the computation of PLs with, however, some significant differences.

1. First, observe that the successive pair-wise iterations are performed with changing power levels (but fixed contingency clearing time); hence, the pre-fault operating conditions have to be refreshed at each iteration; thus, a new power flow must be carried out prior to each stability run.

2. Another difference concerns the choice of initial stability conditions: unlike contingency clearing time, the choice of initial power level is often dictated by practical considerations. Generally, the exploration concerns instabilities caused by contingencies applied under stressed operating conditions. The question is then how much to relax these conditions in order to stabilize the system. Or, conversely, starting with mild conditions, how much to stress them in order to reach stability limit conditions.

---
[14] the machine rotor angles refer to their respective pre-fault rotor angles.

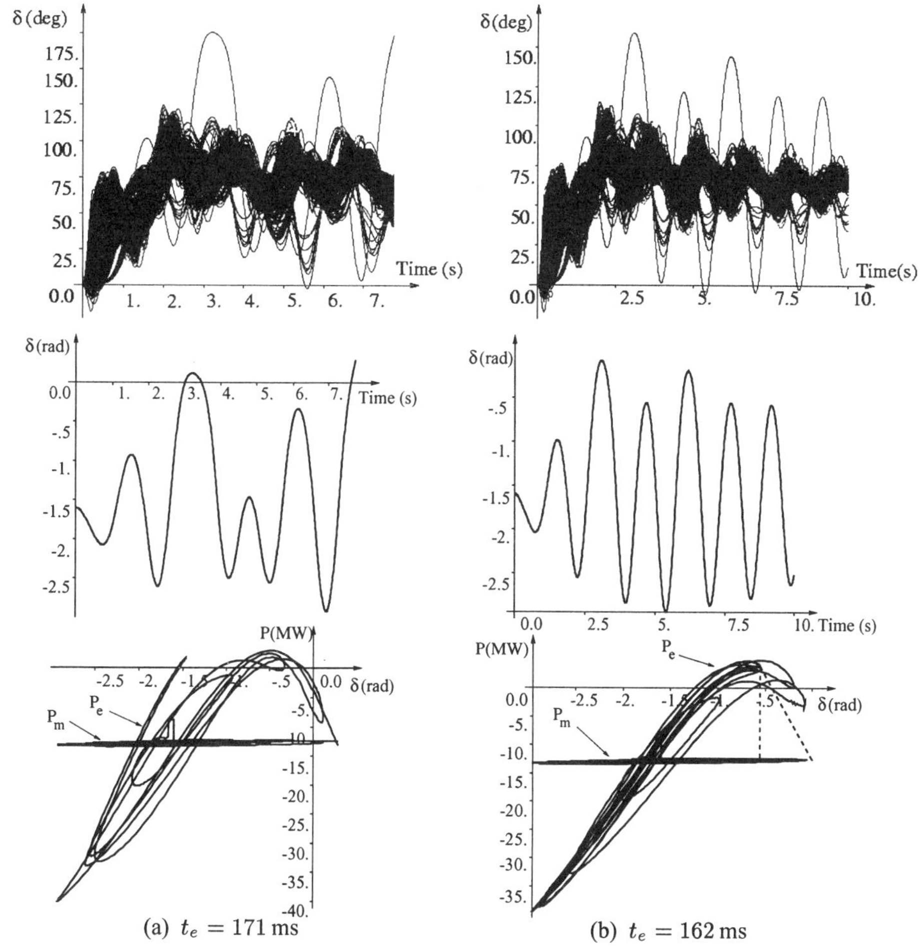

Chapter 4 - PREVENTIVE ANALYSIS AND CONTROL   103

Figure 4.3. Multimachine and OMIB curves for ctg Nr 9 on the 627-machine system.

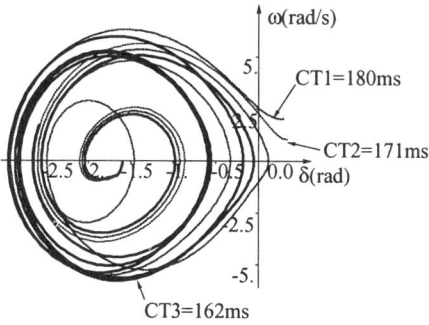

Figure 4.4. Phase plane representation of simulations of Table 4.3. 627-machine system. CCT(SIME) = 168 ms ; CCT(ETMSP) = 168 ms

104   TRANSIENT STABILITY OF POWER SYSTEMS

3. But the main and very important difference is that the search of PLs may be conducted under a large variety of generation and load allocation patterns, yielding a possibly very large number of solutions.

Below we consider two families of approaches. The one consists of changing generation and, to compensate for the generation change, of changing the load level too. The other aims at shifting generation from some machines to other machines in such a way so as to avoid modifying the load (at least to the extent possible).[15]

Let us describe the two approaches successively, then compare and illustrate them.

For reasons discussed below, we will henceforth refer to as "pragmatic"[16] and SIME-based approaches respectively.

2.3.2   "Pragmatic" approach

The approach consists of running successive power flows and stability simulations while adjusting generation and load levels until hitting the stability boundary. The objective may be for example to determine maximum loadability conditions, or maximum power flow on given set of tie-lines. The adjustment may for example be realized by a homothetical variation of generation and consumption.

2.3.3   SIME-based approaches

Obviously, existence of margins as those provided by SIME helps enormously, by guiding the above search.

The approach complies with the following general pattern.

1. Run a stability simulation; compute the margin and the OMIB mechanical power, P_m
2. Decrease P_m by ΔP_m [17], and hence the CMs' generation, P_C, by

$$\Delta P_C = \Delta P_m \text{ (see eq. (3.24))}$$

3. Increase accordingly the NMs' generation, P_N, by

$$\Delta P_N = -\Delta P_C \text{ (see eq. (3.24))}$$

4. Run a power flow[18], and a stability simulation; compute the corresponding new margin and new P_C

---

[15]Throughout the developments of § 2.3 we assume that instabilities are of the upswing type. In case of backswings the reasoning holds valid but with opposite load and generation variations.
[16]"pragmatic" or "heuristic", for want of a more appropriate term; this is why we will use it between inverted commas.
[17]except in the case of backswings, where P_m should be increased.
[18]which readjusts ΔP_N to account for variation of the losses.

5. Extra-(inter-)polate the above two margins, as appropriate
6. Continue according to the pattern of § 2.2.1, suitably adjusted.

Figure 4.5 sketches the beginning of this iterative procedure.

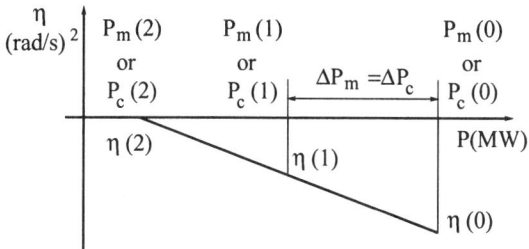

Figure 4.5. Sketch of SIME-based first two steps of PL search

Remark. The choice of ΔP_C may influence computational efficiency[19]. Indeed,

- a too small ΔP_C will increase the number of iterations
- a too large ΔP_C may drive the system to the stable region too far away from its border; the resulting positive margin and/or its linear interpolation with the previous negative margin can be inaccurate, and hence increase implicitly the number of iterations.

Note that after the first two stability runs, the iterations use pair-wise linear extrapolations (together with a final interpolation), and converges readily to the PL sought.

Discussion. Obviously, the above general pattern has a very large number of degrees of freedom with respect to:

- the initial value ΔP_C
- its distribution among CMs (whenever there are many)
- distribution of ΔP_N among NMs

and their combination.
Below, we quote only a few of them.

---

[19] but not accuracy.

2.3.4 Variants of the SIME-based approach

A large number of variants may be obtained by combining the following parameters.

1. **Initial decrease** $\Delta P_m = \Delta P_C$ (above step (2)): it may be chosen as a percentage of ΔP_C (e.g., $0.03 \, P_C$) or computed via the compensation scheme of Fig. 3.12, (see § 3.3.2 of Chapter 3).

2. **Shift of generation from CMs (when there are many).** For example, the total amount ΔP_C may be shifted from CMs :

   (i) proportionally to the nominal power of the respective CM
   (ii) equally from all CMs
   (ii) proportionally to their respective degree of criticalness, i.e. to their electrical distance d_i, defined as the angular deviation of the i-th CM with respect to the most advanced NM, measured at t_u
   (iv) proportionally to their respective inertia, M_i
   (v) proportionally to their respective product, $d_i * M_i$:

$$\frac{\Delta P_{Ck}}{\Delta P_{Ci}} = \frac{M_k d_k}{M_i d_i} . \quad (4.1)$$

The above systematic patterns are only a sample of the combinations one could think of. But they are physically rather sound and easy to implement. Other variants will be described and used in §§ 4.2, 4.3.

3. **Generation reallocation on NMs.** Whatever the pattern used to decrease generation in CMs, the number of degrees of freedom for shifting it to NMs is even larger. However, experience shows that this reallocation does not affect, unless marginally, transient stability. Hence, this opens possibilities to meet other types of objectives, as discussed in Chapter 5.

4. **A final remark:** Stabilizing a case without shedding load cannot be always possible; we should then resort to mixed solutions of generation shifting/shedding combined with load shedding.

2.3.5 Discussion

The number of solutions provided by SIME-based approaches and their combination is undetermined and can be very large.

But which one of these solutions will be optimal ?

Obviously, this question can be answered only after defining the objective sought, and of course, different objectives will call for different solutions.

However, a priori, SIME-based approaches are likely to be superior to "pragmatic" ones. Indeed, decreasing the generation level in CMs while increasing it in NMs brings the two groups of machines closer to each other and hence the system closer to stability; whereas, decreasing generation in CMs *and* NMs has a less beneficial effect. Hence, SIME-based approaches are likely to achieve stabilization much more straightly; besides, they provide many other advantages. An obvious practical advantage is that these approaches require an as small as possible load decrease - if any.

Among other advantages of SIME-based approaches we quote:

- the possibilities to choose various patterns of generation reallocation among CMs
- the possibilities left to NMs to comply with constraints other than transient stability
- the resulting possibilities to set up systematic criteria and make them work automatically, not only for qualitative but also for quantitative assessment and stabilization.

All these important advantages are own to the existence and role of the OMIB. Obviously, T-D methods are just unable to provide such powerful techniques; admittedly, engineering know-how can contribute to shed some light to the "blind" T-D procedures, but only qualitatively and only to a limited extent.

A final remark: since SIME proceeds from "right to left" i.e. from unstable to stable conditions, the PL search may be viewed as a stabilization procedure. (See also observation 1 of § 2.3.8.)

2.3.6 Performances

The accuracy performances of PL search are similar with those of CCT search, discussed in § 2.2.4. But the computing requirements are slightly more demanding, since besides stability simulations, one should perform one power flow per iteration.

2.3.7 Illustration of SIME-based computations

We consider the 3-machine system, apply contingency Nr 2 (which is the most constraining among the 12 selected contingencies, since it has the smallest CCT: 95.7 ms), and compute the PL corresponding to a clearing time of 150 ms. We will use the procedure of § 2.3.3 under the following 3 variants:

- the initial decrease in P_C is chosen rather cautiously: $\Delta P_C = 0.03 \, P_C$; the successive ΔP_C variations are equally shared among CMs
- the initial decrease in P_C is chosen rather cautiously: $\Delta P_C = 0.03 \, P_C$; the successive ΔP_C variations are reallocated proportionally to the product $d * M$ (cf eq. (4.1))
- the initial decrease in P_C is computed by the CS of Fig. 3.12.

The first procedure is described in Table 4.4 and summarized in Fig. 4.6. We will first comment on the synthetic results displayed in this figure before considering the details of the table.

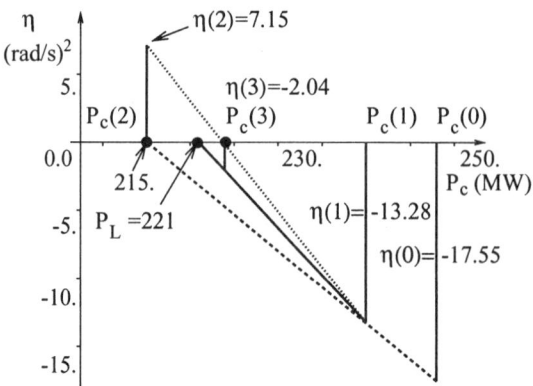

Figure 4.6. Schematic description of the PL search procedure. Contingency Nr 2, cleared at $t_e = 150$ ms

The first iteration is carried out under normal pre-fault operating conditions and contingency clearing time of 150 ms. The OMIB is found to be composed of 2 CMs (m_2 and m_3) and 1 NM (m_1). The pre-fault total power generated by the CMs, P_C, is 248 MW. The corresponding unstable margin is $\eta(248) = -17.55$ (rad/s)^2. It is decided to decrease P_C by about 3 % (actually by 8 MW) and to report these 8 MW on the NM. A power flow is then run to compute the new pre-fault operating conditions.

Using these conditions, the second stability simulation yields a margin $\eta(240) = -13.28$. Extrapolation of the two margins yields 215 MW. As described earlier in § 2.3.1, a new power flow is performed with this value and proper change in P_N.

The third simulation yields a positive margin 7.15. Interpolating this margin with the previous unstable one (-13.28) yields $P_C(3) = 224$ MW ; with this new $P_C(3)$ value and corresponding P_N a new power flow is performed.

The fourth iteration provides $\eta(224) = -2.04$. Since the "distance" ($224 - 215 = 9$ MW) is reasonably small, the algorithm decides to extrapolate the corresponding negative margins (-13.28 and -2.04). The result is $P_C = 221$ MW. The other parameters are computed by a power flow program.

A cross-check of the above result is obtained by computing the CCT of the considered contingency with the new pre-fault conditions provided by the last run of the power flow program: it is found to be exactly 150 ms, i.e., the contingency clearing time imposed above.

Let us now complement the above description with additional information displayed in Table 4.4.

Column 6 shows that $\delta_u = 2.92$ rad corresponds almost to the maximum value of δ_u: $\delta_{u\,max} = \delta_u|_{\eta=0}$. This is in agreement with § 3.3, Fig. 2.5c, of Chapter 2. Incidentally, note that, as suggested by Columns 4 and 6, the slope of the curve δ_u vs P_C is much sharper for large values of P_C than for values near the limit stability conditions.

On the other hand, Column 7 suggests again that the more unstable the case and the shorter the time to instability. Note that the value t_r is provided just for information; actually, the stable simulation has been pursued on the entire simulation period (3 s) to look for possible multiswing phenomena.

Further, Column 2 shows that the mode of instability (CMs) remains unchanged throughout the simulations. In addition, the second part of this column lists the electrical distances of the CMs m_2 and m_3 with respect to machine m_1 at $t = t_u$. These distances "measure" critical

Table 4.4. Detailed description of the PL search of Fig. 4.6 (Ctg Nr 2, cleared at $t_e = 150$ ms). Power rescheduling: equally shared among CMs

1	2		3	4		5		6	7
Iter.	CMs		Normalized margin	Current simul. CM's power		Next simul. CM's power		δ_u (δ_r)	t_u (t_r)
Nr	Name	Electr. dist. (°)	(rad/s)²	Per CM	Total	Per CM	Total	(rad)	(ms)
1	m_2 m_3	158.5 110.9	−17.55	163 85	248	159 81	240	2.44	331
2	m_2 m_3	170.8 117.7	−13.28	159 81	240	146.5 68.5	215	2.68	391
3	m_2 m_3		7.15	146.5 68.5	215	151 73	224	(2.39)	(551)
4	m_2 m_3	182.9 133.3	−2.04	151 73	224	149.5 71.5	**221**	2.92	616

machines' degree of criticalness. Note that machine m_2 remains the most critical throughout the simulations.

Finally, Columns 4 and 5 check that the simulations carried out here have distributed the variations of CMs' power, P_C, (be it increase or decrease) equally in the two CMs. Obviously, this is not the only way of reallocating power among CMs.

In particular, it could be physically more "reasonable" to reallocate power proportionally to the machines' degree of criticalness and size, according to eq. (4.1). The resulting procedure is summarized in Table 4.5.

Finally, Table 4.6 gathers results obtained when the initial ΔP_C value is computed by the compensation scheme.

2.3.8   Observations and comparisons

1.- The above PL searches may be viewed as stabilization procedures, aiming at stabilizing a contingency scenario under given clearing time. Or, conversely, as a means of increasing the contingency CCT (in the above example, from 95.7 ms to 150 ms). These stabilization procedures will therefore be considered as "control" procedures, in Section 4.

2.- Comparing results of Tables 4.4 and 4.5 suggests that reallocating generation on CMs proportionally to their respective product dM allows stabilizing the system at the expense of $248 - 225 = 23$ MW decrease in CMs (and almost

110   TRANSIENT STABILITY OF POWER SYSTEMS

Table 4.5. Search of PL. Contingency Nr 2, cleared at $t_e = 150$ ms. Initial P_C decrease: 3 %. CMs' reallocation according to product dM

1	2		3	4		5		6	7
Iter.	CMs		Normalized margin	Current simul. CM's power		Next simul. CM's power		δ_u (δ_r)	t_u (t_r)
Nr	Name	Electr. dist. (°)	(rad/s)^2	Per CM	Total	Per CM	Total	(rad)	(ms)
1	m_2 m_3	158.5 110.9	−17.55	163 85	248	157 83	240	2.44	331
2	m_2 m_3	169.8 120.1	−12.45	157 83	240	142 78	220	2.68	391
3	m_2 m_3		6.77	142 78	220	147 80	227	(2.40)	(551)
4	m_2 m_3	179.3 137.7	−1.59	147 80	227	145 80	**225**	2.90	626

Table 4.6. Search of PL. Contingency Nr 2, cleared at $t_e = 150$ ms. Initial P_C decrease according to the CS of Fig. 3.12. CMs' reallocation according to product dM

1	2		3	4		5		6	7
Iter.	CMs		Normalized margin	Current simul. CM's power		Next simul. CM's power		δ_u (δ_r)	t_u (t_r)
Nr	Name	Electr. dist. (°)	(rad/s)^2	Per CM	Total	Per CM	Total	(rad)	(ms)
1	m_2 m_3	158.5 110.9	−17.55	163 85	248	147 80	227	2.44	331
2	m_2 m_3	179.3 137.7	−1.59	147 80	227	145 80	225	2.90	626
3	m_2 m_3		0.81	145 80	225	146 80	**226**	(2.73)	(701)

equal increase in NM). This decrease is more important when the generation reallocation is shared equally among CMs (27 MW). This is physically sound.

3.- Comparing Tables 4.4 and 4.5 with 4.6 shows that the initial ΔP_C provided by the compensation scheme of Fig. 3.12 is less cautious and allows "saving" one unstable simulation. Note that when the CS cannot be used (e.g., in case of multiswing phenomena), it is advisable to choose a cautious decrease (e.g., 3%) in order to avoid getting a too stable case in the second step (see Remark of § 2.3.3).

2.3.9 "Pragmatic" vs SIME-based stabilization

Let us now compare the stabilization of the above stability case (3-machine system; contingency Nr 2 cleared at $t_e = 150$ ms) performed by a SIME-based and a "pragmatic" approach.

We specify that the decrease

- in critical machines' generation for the SIME-based approach
- in all the machines' generation and in load for the "pragmatic" approach

are distributed proportionally to their initial generation and/or consumption. We also specify that, here, to facilitate computations of the "pragmatic" approach we use also SIME's margins.

Table 4.7 summarizes the η vs P_C variations for the two approaches and Fig. 4.7 plots them.

Table 4.7. Comparison of "pragmatic" vs SIME-based approach. 3-machine system. Ctg Nr 2; $t_e = 150$ ms.

P_C	η Convent.	SIME-based
248	−17.55	−17.55
242	−15.91	−140.2
236	−13.62	−9.93
231	−11.81	−5.75
225	−9.02	−0.35
219	−6.77	4.34
214	−4.28	7.41
208	−1.12	10.11
203	1.47	12.05

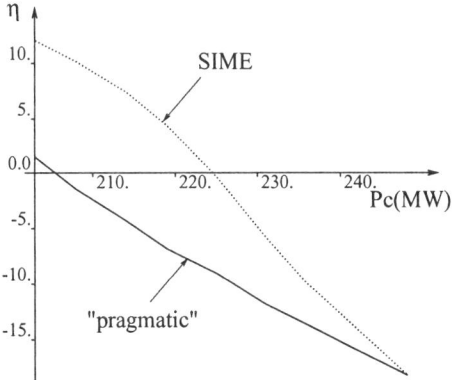

Figure 4.7. "Pragmatic" vs SIME-based stabilization

112   TRANSIENT STABILITY OF POWER SYSTEMS

Observe that the SIME-based stabilization has required:
- generation decrease in CMs of $(248 - 224)$ MW $= 9.7\%$;
- an almost equal increase in the NM;
- and no load decrease.

The "pragmatic" stabilization, on the other hand, requires:
- a decrease of $(248 - 205.5)$ MW , i.e., about 17% decrease in CMs;
- about 17% decrease in the NM;
- and about 17% of load decrease.

In other words, the system stabilization has required:
- by the SIME-based approach: rescheduling 24 MW generated power and no load decrease
- by the "pragmatic" approach: total generation decrease of $42.5 + 72 \times 0.17$ [20] and about as much (54.7 MW) of load shedding.

2.3.10   Concluding remarks

Let us summarize some interesting observations resulting from the PL search of this paragraph.

1. SIME contributes to set up systematic, easy to implement and sound approaches. Hence, it may yield a significant number of sound solutions.
2. Various solutions may encounter various power system specifics and operational objectives. The a priori choice among them is quite transparent.
3. Unlike CCTs, PLs are not well suited as objective indicators for screening contingencies, given the large variety of PL solutions.
4. In the context of Preventive TSA&C, procedures of PL search may be viewed as stabilization, i.e. control procedures. The panoply of approaches developed in this paragraph will therefore be used in Section 4.
5. In short, SIME-based approaches provide interesting techniques for preventive control, sorely missing so far, since T-D methods are helpless in this area.

2.4   Stability limits approximate assessment

2.4.1   Scope and principle

So far we dealt with accurate assessment of stability limits (SLs).

However, accuracy is not always sought. For example, mild contingencies, which do not threaten the system integrity, do not need accurate assessment; rather, what matters is their identification in order to possibly discard them.

SIME provides straightforward identification techniques, relying on approximate SLs and their comparison with a threshold SL.[21]

---

[20] given that the initial generation of the NM (m_1) is 72 MW.
[21] Threshold SLs should be chosen cautiously in order to avoid discarding harmful contingencies; on the other hand, a too cautious choice would fail to discard actually "uninteresting" contingencies, i.e., would produce false alarms.

The proper choice of threshold SLs is an important issue addressed in Section 3. This paragraph deals with approximate search of SLs.

Two types of approximation are used. The one consists of limiting the search to first-swing stability phenomena, thus avoiding lengthy computations on the entire integration period. This approximation uses therefore either negative margins only or, on a stable simulation, stopping criteria described below in § 2.4.4. The second approximation consists of limiting the accuracy of the search. It relies either on previous results of this Section or on compensation schemes developed in Chapter 3. As described below, the former uses a pair of margins, while the latter relies on a single margin.

In all cases the SLs of concern will be CCTs for the reasons advocated so far.

2.4.2 Two-margin approximation

The two-margin approximation relies on the extra-(inter-)polation of two margins corresponding to two successive simulations[22]: of them, the first is necessarily unstable[23] and yields a negative margin, η_0. The second simulation may be stable or unstable:

- if it is stable, the resulting margin is interpolated with η_0 to yield the approximate CCT sought

- if it is unstable, it is stopped at t_u and the resulting margin η_1 is extrapolated with η_0 to yield the approximate CCT sought.

Note that if the approximation seeks for a first-swing CCT, the stable simulation (if any) should be stopped at t_r, instead of being carried out on the whole integration period.

To fix ideas, in the example of § 2.2.4 described in Table 4.1, the approximate CCT of concern is found by extrapolating the margins -44.25 and -34.99 and equals 123 ms. This value is quite close to the actual CCT (131 ms); note however that, here, multiswing phenomena have not been explored, since no stable simulation has been performed on the entire simulation period.

In terms of computing effort, the above approximate assessment requires 0.498 sTDI (0.239+ 0.259), i.e., much less than 4.008 sTDI required for the accurate CCT assessment.

Finally, observe that application of this two-margin approximation on the numerical example of § 2.3.7 provides the approximate value PL = 215 MW (see Fig. 4.6); again, this is quite close to the accurate value (221 MW).

---

[22] Note that the two-margin approximation can also be used to assess PLs, in addition to CCTs.
[23] Remember, if the simulation is too unstable and hence outside the interval of margins' existence, additional simulations (generally two) will be used to recover a margin; however, these very unstable simulations will generally be rather unexpensive.

2.4.3 Single-margin approximation

The single-margin approximation relies on the compensation schemes described in Figs 3.11, § 3.2 of Chapter 3. It is illustrated below, in § 2.4.5. It will be used in the filtering schemes of Chapter 5.

2.4.4 Stopping criteria for first-swing screening

The previous two paragraphs deal with approximate CCT assessment. This paragraph describes a technique which identifies and discards first-swing stable (FSS) contingencies, while assessing and keeping first-swing unstable (FSU) ones. The distinction between FSS and FSU is made on the basis of a threshold clearing time, CT_{thr}.

The procedure is as follows. Upon choosing CT_{thr}, run the T-D program, first in the during-fault then in the post-fault configuration entering at CT_{thr}; continue simulating until reaching one of the following conditions:

(i) the system extreme machines reach a maximum angular deviation *and* a candidate OMIB reaches the stable conditions (2.16)
(ii) a candidate OMIB reaches the unstable conditions (2.15).

In the former case, the contingency is declared FSS and discarded. In the latter case, the contingency is declared FSU and stored for possible finer assessment, along with the corresponding margin and CMs. Note that the candidate OMIB mentioned above is one of the few (about 5) candidates identified along the T-D simulation.

2.4.5 Illustrations on the 3-machine system

We consider the 12 contingencies applied on the 3-machine system (see Section 1 of Appendix B) and assess their CCT according to:

- the approximation of § 2.4.2, using the two clearing times, $t_e = 200$ and 150 ms
- the approximation of § 2.4.3, using the clearing time $t_e = 200$ ms.

Further,

- a contingency which at $t_e = 200$ ms has no margin is declared FSU
- a contingency which meets criteria (i) of § 2.4.4 is declared FSS.

The validity of the above assessments is checked by computing the CCTs provided by SIME. Table 4.8 gathers the results of the investigation.

- Column 2 recalls the accurate CCT values computed by SIME, listed in Column 3 of Table 4.2;
- Column 3 gives the CCT values obtained by the two-margin approximation, where the margins are computed for the following two clearing times: 200 ms and 150 ms.
- Columns 4 list the CCTs relying on the sole margin computed for $t_e = 200$ ms via two of the compensation schemes of Figs 3.11.
- Column 5 identifies the FSU and FSS contingencies for $CT_{thr} = 200$ ms, relying on the CCTs of Column 2.

Chapter 4 - PREVENTIVE ANALYSIS AND CONTROL 115

Table 4.8. Approximate CCT assessment

1	2	3	4		5
Cont. identif.	Accurate CCT (ms)	Two-margin CCT approximation (ms)	Single-margin CCT approximation (ms)		First-swing screening
			t_{c3}	t_{c4}	
cont. 2	95	/	/	/	FSU
3	131	131	88	98	FSU
7	147	148	94	101	FSU
6	189	184	181	181	FSU
1	194	190	194	194	FSU
4	196	195	196	196	FSU
5	220	–	–	–	FSS
10	225	–	–	–	FSS
11	227	–	–	–	FSS
8	251	–	–	–	FSS
12	272	–	–	–	FSS
9	297	–	–	–	FSS

Inspecting the resulting assessment, we observe that:

- in Column 4, t_{c3} and t_{c4} look quite convenient.
- Column 5 shows that all approximate CCTs identify correctly the FSU and FSS contingencies.

3. FILTRA

N.B. The contingency FILTering, Ranking and Assessment (FILTRA) technique developed in this section is mainly transcribed from [Ernst et al., 2000b, Ruiz-Vega et al., 2000b].

3.1 Scope of contingency filtering, ranking, assessment

The previous section has focused on the (accurate or approximate) computation of stability limits (CCTs or PLs) relative to a contingency. But, in practice, the number of "plausible" contingencies is very large; besides, their occurrence is a priori rather unpredictable, apart maybe from a restricted number of contingencies. Yet, the system should be able to withstand any contingency or at least to design "last moment remedial actions" and trigger them, if necessary. On the other hand, a properly designed power system is likely to withstand the large majority of contingencies. Thus, in the context of the preventive mode, to overcome this "curse of dimensionality" we design appropriate contingency filtering tools able to screen out the "mild" and hence "uninteresting" contingencies, while ranking the "potentially interesting ones". These filtering-ranking

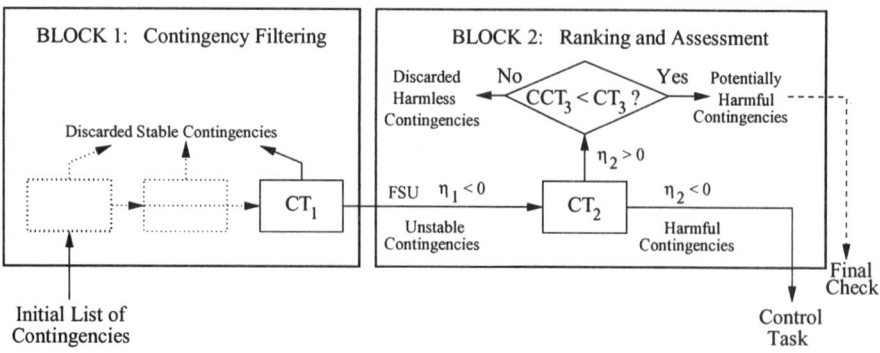

Figure 4.8.   Two-block general organization of FILTRA: $CT_1 > CT_3 > CT_2$

phases should further be followed by an assessment phase able to scrutinize those contingencies whose stability limits have been found to be unacceptably low.

This section deals with the design of a general contingency FILTering, Ranking and Assessment (FILTRA) technique, using tools developed in the previous section.

Section 4 will then set up control techniques able to stabilize the "harmful" contingencies identified and assessed by FILTRA.

FILTRA is a technique designed so as to meet the requirements stated in § 3.2. The resulting general structure is developed in § 3.3 and schematically described in Fig. 4.8. Its mechanism is scrutinized and main properties are highlighted in § 3.4. Paragraph 3.5 proposes illustrative real-world examples. Paragraph 3.6 suggests possible variants. Finally, § 3.7 points out features and draws conclusions.

3.2   Basic concepts and definitions

Any good contingency filter should meet some key requirements, expressed hereafter in terms of conditions. Main terms used in the remainder of the book are also defined.

CONDITION 1: **Classification ability**. A good classifier should be able to screen and rank contingencies on the basis of increasingly severe criteria. According to FILTRA, the various contingencies are classified into first-swing stable or unstable; these latter are then classified into (first- and multi-swing) harmless, potentially harmful or harmful. Further, the harmful ones are ranked according to their degree of severity and assessed. These terms are defined below and illustrated in Fig. 4.8. A contingency is said to be

Chapter 4 - PREVENTIVE ANALYSIS AND CONTROL   117

- *Harmful* (H) if its occurrence drives the system out of step; in other words, a contingency whose critical clearing time is smaller than the assumed time response of system protections;

- *Potentially Harmful* (PH) if it is "almost" harmful, i.e., milder than, but likely to become harmful under slightly modified operating conditions;

- *First-Swing Unstable* (FSU), if under given clearing scenario it drives the system to first-swing instability, *First-Swing Stable* (FSS), otherwise;

- *Harmless* (Hs), if it is FSU[24] but neither harmful nor potentially harmful.

The severity criterion used in the above definitions is contingency clearing time (CT). To classify a contingency as FSS or FSU, the filtering block chooses a first threshold, CT_1,[25] quite larger than the time response of the system protections of concern. Further, to rank an FSU contingency, the second block chooses a second threshold, CT_2 slightly larger than the protections' time response: accordingly, it declares an FSU contingency to be H, if it is unstable for this CT_2; otherwise, it ranks it as Hs or PH depending upon whether it is stable or unstable with respect to a third threshold, CT_3 (see Fig. 4.8).

CONDITION 2: **Accuracy**. The more unstable a contingency, the more interesting from a stability viewpoint, and the more accurately it should be assessed. This is achieved by using detailed power system models to rank and assess the harmful and potentially harmful contingencies.

CONDITION 3: **Reliability**. The contingency filter should be 100 % reliable, i.e., able to *capture all the harmful and potentially harmful contingencies*. This is achieved by the combined use of detailed system models and fairly large threshold values for CT_1 (see also below, § 3.3).

CONDITION 4: **Efficacy**. The contingency filter should have an as low as possible rate of false alarms, i.e., of contingencies suspected by the filter to be harmful or potentially harmful while they are not.

Note that the identification of all harmful and potentially harmful contingencies is a condition of paramount importance for the very validity of the filter, while false alarms may affect its computational efficiency.

CONDITION 5: **Computational efficiency**. The overall procedure of contingency filtering, ranking and assessment should be as fast as possible, whatever the application context. This condition becomes crucial when it comes to real-time operation. The computing requirements of FILTRA are assessed in § 3.4.5.

---

[24] defined for a larger clearing time, $CT_1 > CT_2$
[25] In general, thresholds CT_1, CT_2, should be contingency dependent.

3.3 General design

Figure 4.8 portrays the general two-block structure of FILTRA. The first block is devoted to the filtering task; it may consist of several successive sub-blocks, with increasing modeling details and filtering accuracy, as discussed in § 3.6. The second block ranks and assesses the "interesting" contingencies sent from the first block.

As suggested by Fig. 4.8, the ranking of FSU contingencies relies on margins (η's) computed for two clearing times (CTs), fixed so as to comply with the conditions of § 3.2. Note that CCT_3, shown in the upper part of block 2, is obtained by linear interpolation of η_1 and η_2 (see § 3.4 and sketch (II) of Fig. 4.9). Recall also that CT_3 is an intermediate value between CT_1 and CT_2 ($CT_1 > CT_3 > CT_2$) [26]. Margins and CTs are used as follows.

3.3.1 Contingency filtering block

Whatever the internal structure, the last step of this block performs a stability computation with detailed power system modeling and contingency clearing time CT_1 to classify each contingency as first-swing stable or unstable and, accordingly, to:

- discard the contingency if it is FSS

- send the contingency to block 2 along with its (negative) margin η_1, if it is FSU.

3.3.2 Contingency ranking and assessment block

This block uses detailed power system modeling and a threshold CT_2 to compute a stability margin, η_2 and classify a contingency as:

- Harmful, if $\eta_2 < 0$

- Potentially Harmful (PH) or Harmless (Hs), if $\eta_2 > 0$: the CCT value resulting from the linear interpolation of $\eta_1\ (< 0)$ and $\eta_2\ (> 0)$ decides whether the contingency is unstable for CT_3 ($CCT_3 < CT_3$), and hence PH, or stable ($CCT_3 > CT_3$) and hence Hs.

To summarize, only the harmful contingencies would actually threaten the power system, and deserve finer exploration (see § 3.4.3). The potentially harmful contingencies may be put in a stand-by list and checked after stabilization of the harmful ones.

---

[26] See for example the values used in Fig. 4.9.

3.3.3 Remarks

1.- Detailed power system modeling in the last step of block 1 is used with the twofold objective: for better accuracy, and for allowing extra-(inter-)polating the resulting margin with that computed for CT_2, in block 2.

2.- Many variants of FILTRA may be thought of in order to comply with power system specifics. They all differ in the structure of the filtering block (see § 3.6), whereas the second block, which carries the main properties of the approach, is less liable to changes.

3.4 A particular realization of FILTRA

The general FILTRA technique is here scrutinized on the simple structure of Fig. 4.9, in order to describe its mechanism and uncover main features and properties. This structure will subsequently be used in the simulations of § 3.5; the parameter values and contingency numbers displayed in this figure are borrowed from these simulations.

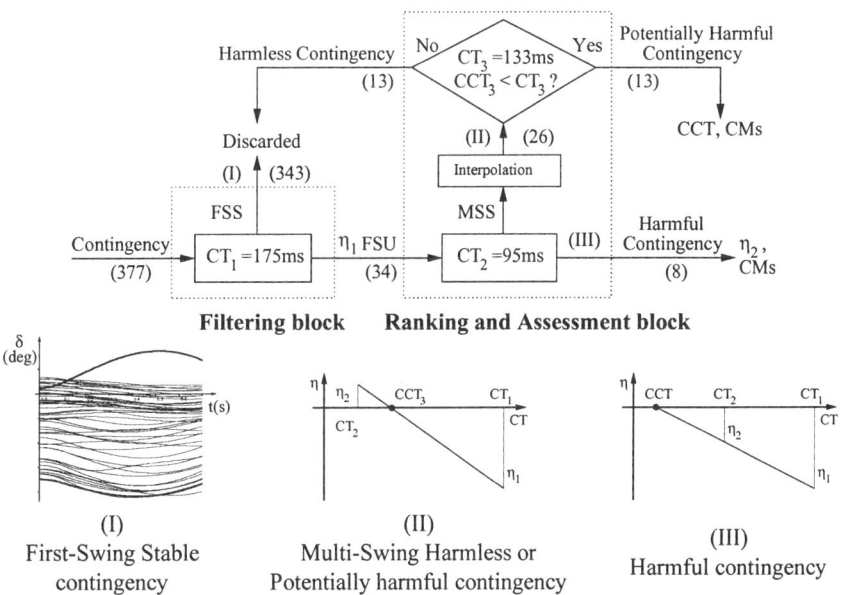

Figure 4.9. A realization of FILTRA. Schematic description of the various contingency classes. Application to the Hydro-Québec power system

3.4.1 Contingency filtering

According to Fig. 4.9, 377 contingencies are inputted to block (I). In order to classify them as first-swing stable or unstable, SIME drives the T-D program, first in the during-fault then in the post-fault configuration entering at $CT_1 = 175\,\text{ms}$. Further, SIME stops the T-D integration as soon as either of the two conditions of § 2.4.4 is met.

In case (i) of § 2.4.4, the contingency is declared to be first-swing stable (FSS) and discarded. In case (ii), the contingency is declared first-swing unstable (FSU) and sent to block 2 along with its corresponding negative margin, η_1, and list of critical machines. For example, in Fig. 4.9, out of the 377 contingencies, 343 are discarded and 34 are sent to the second block.

3.4.2   Contingency ranking

Following the general pattern of § 3.3, SIME ranks the FSU contingencies by driving the T-D program with $CT_2 = 95\,\text{ms}$ onwards. The simulation is either stopped as soon as the instability conditions (2.15) are met or pursued on the entire integration period (5, 10 or 15 s, as appropriate), if the simulation is found to be stable[27]. In the former case, the contingency is declared to be harmful and the corresponding (negative) margin η_2 is computed; in the latter case, the (positive) margin η_2 is computed and interpolated with η_1 to get CCT_3, and: if CCT_3 is larger than CT_3, the contingency is harmless; otherwise, it is potentially harmful and stored in the "waiting list".

3.4.3   Refined ranking of harmful contingencies

To rank harmful contingencies, two parameters obtained as by-products of the above simulations may a priori be considered: the unstable (negative) margin; the time to instability, t_u.

Concerning margins, observe that the approximate CCT obtained by extrapolating the negative margins, η_1, and η_2 (see sketch (III) of Fig. 4.9) might be a good "measure" of contingency severity. However, these two margins (especially margin η_1) seldom exist for very unstable scenarios like those of harmful contingencies. In addition, although normalized, if these margins correspond to different OMIBs they cannot be compared validly.

The time to instability, t_u, seems to be more convenient for ranking contingencies; indeed as already observed, the more unstable a contingency the faster the system loses synchronism. Note, however, that only t_u's referring to the same type of instability may be compared. (See a counter-example and its discussion in § 3.5.) Note that [Ejebe et al., 1997] uses also the time to instability, though computed in a different way.

---

[27] Recall that the reason for performing the time-domain simulation on the entire integration period is to make sure that the contingency is indeed multiswing stable; otherwise, i.e., in case multiswing instabilities are detected, the contingency is harmful and treated as such.

3.4.4 Assessment of harmful contingencies

What mainly characterizes a harmful contingency is its margin, η_2, and corresponding critical machines (CMs). Knowledge of these two pieces of information opens avenues towards control, i.e., stabilization. This issue is addressed in Section 4.

3.4.5 Computing requirements of FILTRA

For the particular FILTRA of Fig. 4.9, the computing times required to classify contingencies into the above four classes are given below, in terms of sTDI:

FSS: $t_r\,(CT_1)$;   H: $t_u\,(CT_1) + t_u\,(CT_2)$;
PH: $t_u\,(CT_1) + MIP$;   Hs: $t_u\,(CT_1) + MIP$.

In the above, $t_r\,(CT_1)$ denotes the time to reach the first-swing stable conditions. Similarly, $t_u\,(CT_1)$ (respectively $t_u\,(CT_2)$) is the time to reach the unstable conditions for CT $=CT_1$ (respectively CT_2). Finally, MIP denotes the "Maximum Integration Period" (e.g., in § 3.5 it is taken equal to 5 s for the EPRI system and 10 s for the Hydro-Québec system).

Note that the refined ranking proposed in § 3.4.3 does not require any additional computing time.

3.4.6 Main properties of FILTRA

Although drawn for the particular FILTRA of Fig. 4.9, the properties uncovered so far and summarized below are quite general.

1. The approach is "unified" and straightforward. Indeed, it uses the same SIME program throughout. Further, the resulting pieces of information are generally used twice: thus, for FSU contingencies, the margin η_1, computed at the filtering block is subsequently used in the second block, together with the margin η_2 to rank these contingencies; similarly, the margin η_2 is subsequently used to assess the severity of H contingencies and, if desired, of PH contingencies as well.
2. The very stable cases are assessed only approximately, thus requiring little CPU time.
3. The more unstable a contingency, and the more detailed and accurate the information provided about it.
4. In all practical simulations, the procedure has shown to be 100 % reliable, i.e., able to capture all harmful contingencies; this reliability is obtained while keeping a low rate of false alarms.
5. The above properties contribute to make the procedure computationally very efficient and compatible with on-line requirements. Besides, the most time-consuming steps may easily be parallelized and performed by distributed computing.

122   TRANSIENT STABILITY OF POWER SYSTEMS

Table 4.9.   Main characteristics of simulated power systems

1	2	3	4	5	6	7
Power system	Nr of buses	Nr of lines	Machine modeling DM	SM	Total power (MW)	Nr of cont.
EPRI A	4112	6091	346	281	76,170	22
EPRI C	434	2357	14	74	350,749	252
Hydro-Québec	661	858	86	0	36,682	377

3.5 Illustrating FILTRA techniques

3.5.1 Simulation description

Two power systems are considered: the EPRI test system C [EPRI, 1995] and the Hydro-Québec (H-Q) power system. These systems are described in Appendix B, while their main characteristics are summarized in Table 4.9. Columns 4 and 5 of the table indicate the number of machines with detailed model (DM) and with simplified model (SM) respectively. On both systems, the contingencies considered are 3-ϕ short-circuits, applied at EHV buses (500 kV for the EPRI and 315, 345 and 735 kV for the H-Q system); they are cleared by tripping one or several lines. Note that the 252 contingencies mentioned in the table for the EPRI system result from the simulation of 36 contingencies under 7 different operating conditions. Paragraph 3.5.4 zooms in simulation results obtained with one of these conditions.

For both power systems, the FILTRA structure and parameters are those displayed in Fig. 4.9. In addition, the number of reported contingencies and their classification correspond to the simulations of the H-Q system.

The T-D program used for the EPRI system is ETMSP [EPRI, 1994], and for the H-Q system is ST-600 [Valette et al., 1987]. These programs are coupled with SIME and, in addition, used as reference for accuracy comparisons. Note that in order to comply with operational uses, the maximum integration period for a stable simulation was fixed at 5 s for the EPRI system and 10 s for the H-Q system.

Below we report on two types of simulations:

- a synthetic assessment of the two-block FILTRA, using the number of contingencies listed in Table 4.9, then a zoom in the assessment of H contingencies (§§ 3.5.2, 3.5.3)
- a detailed classification of the 36 contingencies used with EPRI system C under a given operating point, as provided by the two-block scheme (§ 3.5.4).

3.5.2 Zooming in harmful contingencies

Filtering block.   For the EPRI system, out of the initial list of 252 contingencies, 172 have been found FSS and discarded. The remaining 80 FSU contingencies have been selected and sent to the second block for ranking and assessment.

For the H-Q system, out of the initial list of 377 contingencies, 343 were found to be FSS and 34 FSU.

Table 4.10. Ranking and assessment of harmful contingencies

1	2	3	4	5	6	7
Cont. Nr	η_2 (rad/s)^2	Nr of CMs	P_C (MW)	$(t_{u1})\, t_{u2}$ (ms)	Rank	CCT (ms)
EPRI 88-machine test system C							
1	*	6	4,832	(485) 395	1	0
30	−1.20	28	20,008	(550) 1010	2	0
11	−0.81	37	26,938	(365) 1325	3	66
10	−0.70	39	27,714	(390) 1395	4	70
Hydro-Québec system							
33	*	5	5572	(258) 320	1	0
39	*	5	5572	(258) 322	2	0
1405	*	5	5572	(284) 364	3	0
13	*	5	5572	(258) 322	4	0
1405	−5.41	5	5572	(332) 438	5	48
243	−0.85	8	2293	(324) 971*	6	67
37	−4.73	5	5572	(260) 414	7	80
636	−1.74	1	42	(258) 921	8	92

Ranking and assessment block. For the EPRI system: the 80 FSU contingencies are decomposed into 31 Hs, 25 PH and 24 H contingencies. For the H-Q system: the 34 FSU contingencies are decomposed into 13 Hs, 13 PH and 8 H contingencies.

The harmful contingencies are further ranked, according to § 3.4.3. The obtained results are gathered in Table 4.10, where:

- Column 2 gives the margin η_2 computed for CT= 95 ms. An asterisk indicates that there is no margin; (P_a remains always positive). Obviously such cases are very unstable;
- Column 3 specifies the number of critical machines;
- Column 4 lists the total power generated by these CMs. This information is useful, though not indispensable;
- Column 5 lists the time to instability; the first, between brackets, refers to the first simulation, using $CT_1 = 175$ ms; the second to the second simulation, using $CT_2 = 95$ ms. Note that all harmful contingencies are first-swing unstable, apart from contingency Nr 243 of the H-Q system which loses synchronism after a backswing excursion;
- Column 6 ranks the contingencies in increasing order of t_{u2} (apart from contingency 243, which has a different mode of instability);
- Column 7 provides the reference CCTs furnished by SIME.

Finally, we mention that for the EPRI system only the 4 harmful contingencies corresponding to operating point Nr 6 (see Section 3 in Appendix B) are displayed, the others exhibiting quite similar behavior.

Comments. 1.- Concerning contingency ranking, the margin is not a good indicator; in particular, because most of the harmful contingencies do not have margin for $CT_2 = 95$ ms, and a fortiori for $CT_1 = 175$ ms.

124   TRANSIENT STABILITY OF POWER SYSTEMS

In contrast, the time to instability shows to be a convenient contingency severity indicator; the ranking of column 6 of Table 4.10 coincides with that relying on the reference CCTs of last column, except for contingency Nr 243, which has a different mode of instability: it loses synchronism after a backswing excursion.

2.- The distribution of contingencies into FSS, Hs, PH and H is much more realistic for the H-Q than for the EPRI system, where the 252 contingencies under consideration seem to result from a pre-selection having discarded most of the stable contingencies.

3.5.3   Performances

Reliability. Simulations not reported here show that all contingencies discarded by the filtering block are indeed stable, and that all harmful contingencies are properly captured. Note also that the CCTs obtained with the T-D programs run alone are found in perfect agreement with the CCTs of SIME [Ruiz-Vega et al., 2000b]. This is corroborated by the results listed below, in Table 4.11.

Computational efficiency. The only lengthy computation is the one concerning harmless contingencies: 13 out of the 377 for the H-Q system, and 13 out of the 252 for the EPRI system. However, these computations are worth for guaranteeing full reliability.

Ranking ability of FILTRA. It is very good, according to the comparison of Columns 6 and 7 of Table 4.10.

Computational performances. In terms of sTDI, the total computing times required by FILTRA for the H-Q system are as follows (see also § 3.4.5).

$$
\begin{array}{llll}
\text{FSS} & : & t_r \text{ (CT1)} & = & 141.6 \\
\text{H} & : & t_u(\text{CT1}) + t_u(\text{CT2}) = 2.3 + 4.1 & = & 6.4 \\
\text{PH} & : & t_u(\text{CT1}) + \text{MIP} = 5.5 + 130 & = & 135.5 \\
\text{Hs} & : & t_u(\text{CT1}) + \text{MIP} = 5.9 + 130 & = & 135.9 \\
& & \text{Total time} & = & 419.4 \quad \text{sTDI}
\end{array}
$$

This total may be decomposed into the time required by:

- the first block: 155.3 sTDI
- the second block: 264.1 sTDI.

Of the above 264.1 sTDI, 260 sTDI are spent to run 26 stable simulations on the entire integration period (10 sTDI per contingency), while the harmful contingencies require only a few percentage (about 1.5%). In other words, apart from the contingency filtering of the first block, which is unavoidable, most of the computing time is spent to explore existence of multiswing phenomena. This computation might be avoided if such phenomena are not of concern (for example, if the system operator knows by experience that they don't exist).

In this latter case, the time needed to compute the H, PH and the Hs contingencies reduces to about 30.5 sTDI, and the total computing effort from 419.4 to 172.0 sTDI. The corresponding mean computing times are, respectively, 1.1 and 0.5 sTDI per contingency.

3.5.4 Zooming in classification ability of FILTRA

Table 4.11 zooms in one of the 7 operating conditions of the EPRI system mentioned earlier in § 3.5.2, in order to assess reliability and computational requirements of FILTRA; the accuracy of SIME with respect to ETMSP program run above is also compared.

The table is subdivided in three blocks:

- the first gathers the benchmark CCTs obtained by the ETMSP program run alone;
- the second gives the CCTs obtained by SIME and their comparison with the above; more specifically, Columns 5 and 6 display respectively the difference in ms and in percentage between the CCT values computed by the ETMSP alone and by SIME. The accuracy is assessed in ms and %:

$$\Delta(CCT) = CCT(ETMSP) - CCT(SIME) \quad (ms)$$
$$\Delta(CCT)(\%) = [\Delta CCT / CCT(ETMSP)] \times 100 \quad (\%)$$

- the third block reports on results obtained with FILTRA.

Obviously, SIME is in a good agreement with ETMSP: apart from few exceptions, the discrepancies are within the tolerance range of ETMSP and SIME (± 4 ms). Note that this goes along the general observation: the agreement of SIME with the T-D program that it uses has been obtained consistently, whatever this program, the power system, and its modeling. Actually, SIME gives an even more reliable and unbiased assessment than the T-D program.[28]

Regarding FILTRA, the results are gathered in Columns 7 to 9 of the table:

- Column 7 shows the number of simulations required;
- Column 8 displays the corresponding sTDIs;[29]
- Column 9 gives the contingency classification under the conditions specified so far.

Comparing the classification of Column 9 with the actual CCT (Columns 2 to 4), one observes that FILTRA has provided consistently reliable results: it has captured all H contingencies and, in addition, has classified correctly the remaining contingencies into FSS, PH and Hs.

3.6 Variants of the filtering block

The ultimate objective of the filtering block of Fig. 4.8 is to realize a good compromise between reliability (ability to capture all the harmful contingencies), efficacy (as low as possible rate of false alarms) and computational efficiency. Note that accuracy is not the main concern at this stage. Hence, many approximate filtering schemes may be thought of, a sample of which are described below.

An issue of concern is whether and to which extent it is interesting to use power system simplified modeling (SM). Actually, this raises the twofold question: (i) is the SM at all usable ? (ii) if yes, does SM give a "reasonable"

---

[28] SIME's (in)stability criteria are unambiguously defined, while those of a T-D program are system- and operator-dependent, see § 2.4.4 of Chapter 2

[29] Note that the resulting mean filtering time is about 2.2 sTDI, i.e., larger than the one mentioned earlier, in § 3.5.3. This is because the dynamics of the considered EPRI test system is slower than that of the Hydro-Québec system.

Table 4.11. Simulation result with EPRI test system C. Case # 6. Adapted from [Ruiz-Vega et al., 2000b]

1	2	3	4	5	6	7	8	9
	ETMSP			SIME			FILTRA	
Cont. Nr	CCT (ms)	Rank	Cont. (ms)	Δ(CCT) (ms)	Δ(CCT) (%)	Nr of Sim.	sTDI	class
1	0	1	0	0	0.0	2	0.88	H
2	113	5	115	-2	-1.8	2	5.7	PH
3	156	11	161	-5	-3.2	2	6.36	Hs
4	145	10	147	-2	-1.4	2	6.32	Hs
5	172	16	179	-7	-4.1	1	1.49	FSS
6	277	22	280	-3	-1.1	1	1.15	FSS
7	320	25	316	4	1.3	1	1.13	FSS
8	430	32	426	4	0.9	1	1.09	FSS
9	297	23	308	-11	-3.7	1	1.13	FSS
10	70	4	72	-2	-2.9	2	1.79	H
11	66	3	69	-3	-4.5	2	1.69	H
12	172	17	174	-2	-1.2	1	1.51	FSS
13	172	18	174	-2	-1.2	1	1.51	FSS
14	168	12	173	-5	-2.9	1	1.64	FSS
15	168	13	173	-5	-2.9	1	1.64	FSS
16	168	14	173	-5	-2.9	1	1.64	FSS
17	168	15	173	-5	-2.9	1	1.64	FSS
18	316	24	324	-8	-2.5	1	1.13	FSS
19	324	26	325	-1	-0.3	1	1.12	FSS
20	434	33	436	-2	-0.5	1	1.08	FSS
21	434	34	436	-2	-0.5	1	1.08	FSS
22	113	6	116	-3	-2.7	2	5.72	PH
23	113	7	116	-3	-2.7	2	5.72	PH
24	172	19	174	-2	-1.2	1	1.51	FSS
25	172	20	174	-2	-1.2	1	1.51	FSS
26	328	27	331	-3	-0.9	1	1.12	FSS
27	328	28	331	-3	-0.9	1	1.12	FSS
28	434	35	436	-2	-0.5	1	1.08	FSS
29	438	36	463	-25	-5.7	1	1.08	FSS
30	0	2	0	0	0.0	2	1.56	H
31	328	29	332	-4	-1.2	1	1.14	FSS
32	328	30	332	-4	-0.2	1	1.14	FSS
33	332	31	333	-1	-0.3	1	1.14	FSS
34	215	21	218	-3	-1.4	1	1.19	FSS
35	137	8	141	-4	-2.9	2	5.9	Hs
36	137	9	141	-4	-2.9	2	5.9	Hs

(or realistic) account of the system behavior, i.e. of the system modeled in its normal (detailed) way ? This twofold question may receive many answers:

- "no": the real system modeling is so sophisticated that SM is meaningless;
- "yes", but: SM does give a picture of the real power system behavior but also introduces distortions (e.g., multiswing phenomena which may disappear with detailed models (DM));
- "yes, indeed": the power system behaves in a similar way with SM and DM, though its transient stability limits (power limits or critical clearing times) are generally lower with SM than with DM.

Obviously, systems belonging to the third class are good candidates for a pre-filter with SM, whereas those belonging to the first class cannot use such SM filters. The "yes, but" category is more difficult to apprehend and needs off-line tuning on the considered power system. For example, if multiswing phenomena exist, one should determine the maximum ratio between first and multiswing CCTs, in order to fix a first CT small enough to avoid missing any multiswing instability, but large enough to avoid unduly large number of false alarms.

The above considerations lead to three ways of screening contingencies; (i) first-swing (in)stability; (ii) approximate CCTs relying on a single-margin; (iii) approximate CCTs relying on two margins.

Filter (i) was described in § 2.4.4 and used in § 3.4, Fig. 4.9. Filter (ii) was described in § 2.4.3. It will be used in Chapter 5. Filter (iii), described in § 2.4.2, may be designed so as to detect multiswing phenomena; in this case, to save CPU, it is advised to use it only when simplified modeling is practicable.

The above elementary classifiers may further be combined to yield a large variety of filters, especially when SM is of concern. The choice must take into account power system specifics and could hardly be discussed on general grounds. The need for off-line tuning appears clearly, notably, in order to inform about the interest in using SM, the existence and nature of multi-swing phenomena, and the choice of threshold CTs. Note that this tuning should be performed once and for all, unless the power system undergoes major changes.

3.7 Concluding remarks

FILTRA is a general approach to contingency filtering, ranking and assessment. It is made up of two blocks, one for screening contingencies, the other for ranking and assessing the "potentially interesting" ones.

The approach is unified, accurate, flexible and powerful: unified, since it uses the same transient stability package throughout and takes multiple advantages of each computed margin; accurate, since it achieves a faithful assessment of the T-D program; flexible, since it may handle any power system modeling, contingency scenario and mode of (in)stability; powerful, since it furnishes efficient filtering, ranking and assessment tools.

From the general two-block structure of Fig. 4.8, the particular FILTRA scheme of Fig. 4.9 was designed to comply with the specifics of the two power systems used for illustration.[30] Throughout, the power systems were simulated with detailed modeling. The technique was shown to be reliable (i.e. to capture all harmful contingencies), accurate and computationally efficient, i.e., compatible with real-time requirements under detailed power system modeling.

4. PREVENTIVE CONTROL

4.1 Generalities

Preventive control aims at stabilizing cases involving anticipated contingencies whose occurrence would cause system's loss of synchronism. For convenience, we will say in short that control deals with harmful contingencies, and aims at designing countermeasures or corrective actions able to "stabilize" them. The decision about whether to actually trigger such actions is left to the operator.

In this section the corrective actions rely on generation shifting and rescheduling considered earlier in §§ 2.3.3, 2.3.4 of this chapter. The contingencies, supposed to be assessed by their margin and CMs, will be stabilized by canceling out this margin. As described in § 2.3.3, the procedure is iterative, and may be initialized either by using the compensation scheme of Fig. 3.12 or by applying an arbitrary (but suitable) initial OMIB generation change. This change may be reported on the individual system machines in very many ways, some of which have been identified in § 2.3.4.

On the other hand, whenever there are more than one harmful contingencies, one may wish to stabilize them *simultaneously*.

The following paragraphs consider in a sequence individual and simultaneous stabilization of contingencies.

4.2 Single contingency stabilization

4.2.1 Principle of generation reallocation

For a single unstable contingency, the stabilization procedure of §§ 2.3.3, 2.3.4 consists of computing the corresponding (negative) margin and from there on ΔP_C and ΔP_N defined by (3.21) to (3.24). The next step concerns

---

[30] As an indication, recall that over 600 contingencies have been screened; of them, about 82 % were readily discarded by the filtering block, while the others were classified into harmless (7 %), potentially harmful (6 %), and harmful (5 %) contingencies. These latter were further ranked in terms of severity and assessed in terms of their margin and critical machines.

generation rescheduling of CMs (distribution of total generation shifting of CMs whenever there are many), and of NMs. Recall that the initial ΔP is only approximate, be it provided by the compensation scheme of Chapter 3 or taken as a percentage of the initial P_C. Hence, the procedure will be iterative.

Experience shows and engineering common sense suggests that the efficiency of this iterative procedure depends only marginally on the choice of NMs on which to distribute ΔP_N.[31] But it depends significantly on the way ΔP_C is reallocated among CMs. A convenient way suggested in § 2.3.4 consists of distributing ΔP_C according to the degree of the CMs criticalness, cf. eq. (4.1).

4.2.2 Illustration

This procedure has already been illustrated on the 3-machine system (see § 2.3.7 and Tables 4.4 to 4.6).

Below, we illustrate it on the EPRI and Hydro-Québec systems simulated in § 3.5, where FILTRA identified the harmful contingencies listed in Table 4.10.

The iterative procedure starts using ΔP_C equal to $P_C - \frac{P_C}{(1.03 \text{ or } 1.1)}$, depending upon whether an initial margin exists or not.

Table 4.12 summarizes the results obtained with four dangerous contingencies, two for each power system (one with, the other without initial margin). Column 2 of the table provides the margin values; in their absence, the asterisk indicates that, instead, the "minimum distance", $P_{a\,min}$, between P_m and P_e curves is given (in MW). In Column 4, P_{Ci} denotes the initial power of the CMs, i.e., the power for which the stability margin of Column 2 was computed at the iteration of concern; in Column 5, ΔP_C denotes the suggested change in P_{Ci}. In Column 6, P_{Cf} denotes the final value of P_C; this value is used as the initial P_C value for the next iteration (provided that the critical group does not change from one iteration to the other).

The stabilization procedure starts (iteration Nr 0) with the output data of the second block of FILTRA reported in Table 4.12: η_2 (or asterisk), number of CMs, and P_C.

To facilitate explanations, let us comment on the case Nr 13 of the H-Q system, where the group of CMs is the same for all successive simulations. A first iteration is run using $\Delta P_C = (5572 - 5572/1.1) = 507 \text{ MW}$. This ΔP_C decrease is distributed among CM, and an increase of the same amount is distributed among non-critical machines. A load flow is then run, followed by a transient stability simulation. The results are shown in the table: the procedure converges after three iterations; the power of the group of CMs guaranteeing stabilization is finally found to be 4791 MW (in bold in the table); in other words, stabilizing this case implies shifting 14 % of the CMs' initial generation power.[32] This is admittedly a quite large decrease; it reflects the severity of this stability case.

The same procedure yields the power limits for the other cases in Table 4.12, as well. Observe that, generally, cases which involve CM changes during the procedure and/or very unstable

---

[31] Though it may be of great importance for other issues such as maximum power transfer or cost, see Chapter 5.
[32] The corresponding generation reallocation in NMs is here proportional to their initial power.

Table 4.12. Stabilizing harmful contingencies individually

1	2	3	4	5	6	7
Iter. Nr	η (rad/s)^2	Nr of CMs	P_{Ci} (MW)	$-\Delta P_C$ (MW)	P_{Cf} (MW)	t_u (ms)
EPRI test system						
Cont.1						
0	−12.55*	6	4831	−439	4392	395
1	−10.30*	7	5162	Extr.	3448	460
2	−1.74*	7	3448	Extr.	2826	745
3	−2.83	7	2826	−254	2572	1040
4	−0.90	7	2572	Extr.	2454	1320
5	0.37	7	2454	Inter.	**2489**	5000
Cont.30						
0	−1.20	28	20008	−583	19425	1010
1	−0.31	16	9175	−268	8907	1625
2	0.0	39	**26889**	–	–	3025
Hydro-Québec system						
Cont.13						
0	−2.01*	5	5572	−507	5065	364
1	−4.53	5	5065	−148	4917	470
2	−2.12	5	4917	Extr.	4787	670
3	0.07	5	4787	Inter.	**4791**	10000
Cont.37						
0	−4.73	5	5572	−163	5409	414
1	−1.97	5	5409	Extr.	5292	556
2	1.1	5	5292	Inter.	**5334**	2000

behavior (without initial margin) require a larger number of simulations; nevertheless, this number remains reasonably small (see contingency Nr 1 of the EPRI system which accumulates the two "difficulties").

4.2.3 Comparing stabilization patterns

Paragraph 2.3.4 proposed a sample of possible patterns for stabilizing contingencies. Of course, there are many others. Below, we report on comparative simulations performed on the Brazilian power system described in Section 4 of Appendix B. Three different patterns are compared, namely:

Pattern #1: stabilization by reallocating power on the CM Nr 2707 only
Pattern #2: stabilization by reallocating power on the CM Nr 2706 only

Table 4.13. Comparing stabilization patterns on the Brazilian power system

1	2	3	4	5		6		7	
Cont.	Identif.	Initial	P_{Ci}	Pattern #1		Pattern #2		Pattern #3	
Nr	CMs	d_i (°)	(MW)	ΔP	P_{Cf}	ΔP	P_{Cf}	ΔP	P_{Cf}
687	2707 2706	53 41	350 250 (600)	−72	278 250 (528)	−109	350 141 (491)	−77	304 219 (523)
656	2707 2706 2705 2704	112 101 90 56	350 250 122 100 (822)	−71	279 250 122 100 (751)	−85	350 165 122 100 (737)	−71	319 226 111 95 (751)
686	2707 2706	53 48	350 250 (600)	−39	278 250 (561)	−51	350 199 (549)	−42	328 230 (558)
692	2706 2705 2707	43 40 38	250 122 350 (722)	−19	250 122 331 (703)	−19	231 122 350 (703)	−19	243 118 342 (703)
690	2705 2706 2707	46 41 27	122 250 350 (722)	−10	122 240 350 (712)	−10	122 250 340 (712)	−10	119 246 347 (712)

Pattern #3 : stabilization by reallocating power on all CMs, according to eq. (4.1).

Table 4.13 collects these results for five harmful contingencies, identified in Column 1.

Column 2 identifies the CMs.

Column 3 lists their corresponding initial electrical distance d.[33]

Column 4 lists their corresponding initial generation (P_{Ci}) and, between brackets the corresponding sum.

Columns 5 to 7 collect the results relative to the three rescheduling patterns; ΔP denotes the total generation decrease, and P_{Cf} the corresponding generation of each CM (and their sum).

Comparing the three patterns in terms of total generation decrease in CMs, ΔP, we observe that:

---

[33] Remember, d_i is the angular deviation between the ith CM and the most advanced NM, computed at t_u.

132   TRANSIENT STABILITY OF POWER SYSTEMS

- all three patterns are equally good for contingencies Nr 692, 690
- Pattern # 1
  - is more interesting (i.e., requires smaller decrease ΔP) for contingency Nr 687 and 686;
  - is equally good with Pattern # 3 for contingency Nr 656
- Pattern # 2 seems less effective, probably because for severe contingencies machine Nr 2706 is not the most critical
- Pattern # 3 is never "the best" but provides consistently good results.

Even if the above comparison cannot help draw conclusions, it shows however that different patterns may yield quite different practical results.

Incidentally, note that the number of iterations needed to stabilize the various cases and patterns is 2 or 3: [34] obviously, the number of iterations has not increased with the system size (compare with Tables 4.4 to 4.6, corresponding to the 3-machine system).

4.3   Multi-contingency simultaneous stabilization

4.3.1   Principle of generation reallocation

To stabilize the whole set of harmful contingencies *simultaneously*, the procedure of § 4.2.1 may readily be adapted as follows.

(i) For a given unstable contingency: compute the decrease in generation of the group of CMs following the procedure of § 4.2.1 and

- if this group contains a single CM: impose this decrease on that CM
- if this group contains many CMs, distribute this decrease according to eq. (4.1).

(ii) Proceed similarly with all unstable contingencies.

(iii) For each CM, choose the generation decrease to be the maximum among those computed in above steps (i), (ii);

(iv) Compute the total generation decrease obtained for all CMs, and compensate by an equal generation increase in NMs.

(v) Run a power flow program, followed by a T-D program to assess the new margins of all contingencies, and decide whether to iterate further or not.

4.3.2   Illustration on the 3-machine system

We consider again the 3-machine system for which contingencies Nr 2, 3 and 7 are found to be harmful when cleared at 150 ms.[35]

Table 4.14 summarizes the obtained results.

---

[34] More about the involved computations may be found in [Ruiz-Vega et al., 1998]

[35] This value was chosen quite large on purpose, i.e., in order to get many harmful contingencies. Further, the example is a little artificial, since the CMs are necessarily almost always the same. Nevertheless, it is interesting for illustration.

Table 4.14. Simultaneous stabilization procedure. 3-machine system. Contingencies cleared at $t_e = 150$ ms

1	2	3	4	5	6	7	8
Cont. Ident.	η (rad/s)^2	Electr. dist. d (°)	CMs	P_{Ci} (MW)	$-\Delta P_C$ (MW)	P_{Cf}	t_u (t_r) (ms)
			Iteration Nr 1				
Ctg 2	−17.55	158.5 110.9	m_2 m_3	163 85 248	−8	157 83 240	331
Ctg 3	− 9.98	101.3	m_2	163	−5	158	381
Ctg 7	−0.84	162.8 145.0	m_2 m_3	163 85 248	−8	**157** **83** 240	616
			Iteration Nr 2				
Ctg 2	−12.45	169.8 120.1	m_2 m_3	157 83 240	−20	**142** **78** 220	391
Ctg 3	−2.96	131.1	m_2	157	−3	154	481
Ctg 7	4.28	162.8 145.0	m_2 m_3	157 83 240	+9	162 85 249	(469)
			Iteration Nr 3				
Ctg 2	6.77	169.8 120.1	m_2 m_3	142 78 220	+9	**147** **80** 227	(551)
Ctg 3	10.92	131.1	m_2	142	+12	154	(391)

Observe that, of course, to stabilize the most constraining contingency (Nr 2), the other two (especially contingency Nr 7) have to be overstabilized. Note also that, again, the power decrease in CMs has been compensated by a generation increase in the NM, thus avoiding any load shedding.

And again, the procedure behaves almost the same way, be it applied to a large system or to an academic small system. This is illustrated with the example reported below.

4.3.3 Illustration on the Brazilian system

This example uses again the Brazilian system, for which 14 harmful contingencies have been identified, when cleared at $t_e = 167$ ms (about 10 cycles). Table 4.15 summarizes the results.

Column 3 of the table lists the power generation change (generally decrease) in CMs. Note that the initial change, ΔP_1, is computed via the compensation scheme of Fig. 3.12.

Column 4 identifies the CMs, and lists between brackets, the corresponding electrical distance and generation decrease reported in each CM according to eq. (4.1). The bold face characters denote the CM and corresponding generation decrease chosen (i.e., the maximum power decrease). Table 4.15 shows that the stabilization procedure needs two iterations only, except maybe for contingency Nr 672 for which an additional generation decrease might be needed.

134   TRANSIENT STABILITY OF POWER SYSTEMS

Table 4.15. Simultaneous stabilization procedure. Brazilian 56-machine system. Contingency cleared at $t_e = 167$ ms. Adapted from [Bettiol et al., 1999a]

1	2	3	4	5	6
	\multicolumn{3}{c}{Iteration #1}	\multicolumn{2}{c}{Iteration #2}	
Cont. Nr	η_1 (rad/s)^2	$-\Delta P_{C1}$ (MW)	CM$_i$ (d_i ; $\Delta P_{Ci} XS$)	η_2 (rad/s)^2	$-\Delta P_{C2}$ (MW)
687	−23.9	−70	2707(53; −42); 2706(41; −28);	1.4	4
656	−17.1	−71	2707(112; −31); 2706(101; −24); **2705(90; −11); 2704(56; −5)**;	3.1	14
686	−12.4	−38	2707(53; −20); 2706(48; −18);	7.7	29
672	−14.7	−82	**2712(89; −82)**;	−0.7	−4
630	−15.6	−45	**2707(31; −45)**	8.9	16
680	−9.7	−52	2712(94; −52)	4.3	25
708	−9.4	−50	2712(96; −50)	4.0	25
598	−9.4	−50	2712(95; −50)	4.1	25
629	−16.5	−31	**2706(21; −31)**	13.9	14
463	−8.0	−43	2712(94; −43)	5.8	34
692	−4.9	−19	2706(43; −7); 2705(40; −4); 2707(38; −9);	15.4	44
707	−6.7	−35	2712(95; −35)	7.2	42
690	−2.4	−10	2705(46; −3); 2706(41; −4); 2707(27; −3);	14.8	75
437	−1.1	−2	**2573(103; −2)**	1.7	1

However, the small size of the corresponding margin (−0.7) suggests that this decrease would be insignificant.

Note that stabilizing the 14 contingencies simultaneously has required 180 MW, i.e. 8.5 % of the total CMs' initial generation power. Again, this power decrease in CMs has been compensated by increasing generation in NMs, without any load shedding.

4.3.4   Stabilizing inter-area mode oscillations

Description.   Typically, inter-area oscillations involve a large group of machines swinging against another group of machines.

To stabilize a contingency creating inter-area oscillations, it seems a priori convenient to act on a sub-set of CMs, generally chosen among the more advanced ones, rather than the whole set of CMs.

To stabilize *simultaneously* many contingencies creating inter-area oscillations it seems reasonable to choose CMs common to the various contingencies,

Chapter 4 - PREVENTIVE ANALYSIS AND CONTROL 135

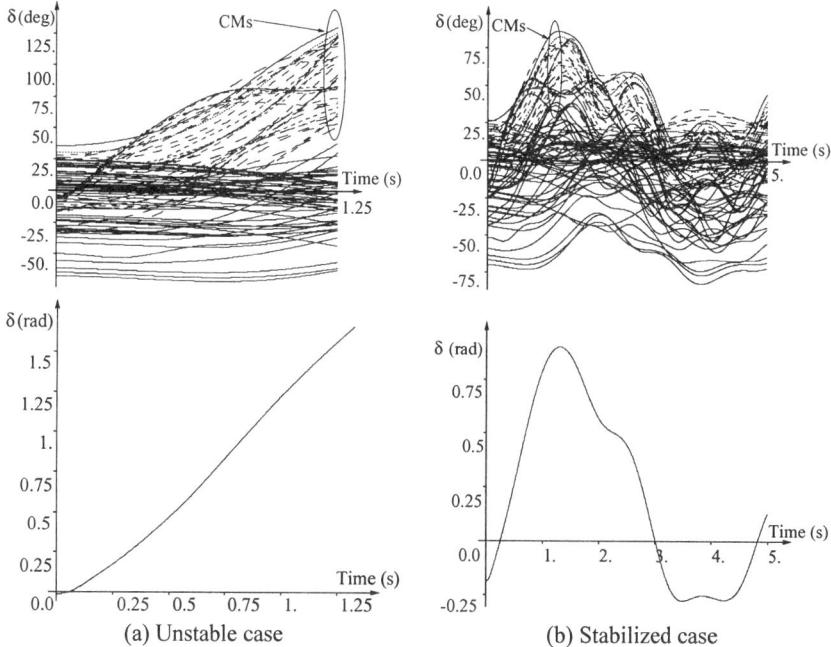

Figure 4.10. Multimachine and corresponding OMIB swing curves of an inter-area mode. EPRI 88-machine test system. Contingency Nr 11; $t_e = 95$ ms

and more specifically those common CMs which are the more advanced ones for the more harmful contingencies.

Illustration. We illustrate these matters on the EPRI 88-machine system on which four contingencies create inter-area oscillations, involving 33 to 38 CMs.[36]

One of these inter-area oscillations is plotted in Fig. 4.10. The simultaneous stabilization procedure is described in Table 4.16. This stabilization takes advantage of the fact that at least 10 CMs are common to all 4 contingencies (actually, there are 32 common CMs).

Column 1 of Table 4.16 lists the 10 more advanced CMs for contingency 1, as well as the initial generation of these machines. On the other hand, Columns 2 to 5 of the table specify these machines' classification for the other contingencies, along with their respective angular distance (in degrees), d_i, from the corresponding most advanced NM. Note that of these 10 CMs, 7 are top machines for contingencies Nr 1 and 30, which are the severer contingencies

---

[36] With respect to the contingencies listed in Table 4.10, contingency Nr 1 has been slightly modified so as to get a margin. The operating conditions are also slightly different.

Table 4.16. Simultaneous stabilization of multi-contingency inter-area mode oscillations. Adapted from [Ruiz-Vega et al., 2000b]

1		2	3	4	5
		Cont. 1	Cont. 10	Cont. 11	Cont. 30
Iteration # 0					
η_0 (rad/s)^2		-0.88	-0.96	-0.83	-1.21
P_{C0}		24623	26162	26162	27014
Nr of CMs		32	36	36	38
Machines		Angle order	Angle order	Angle order	Angle order
Nr	P_{Ci}	(d_i)	(d_i)	(d_i)	(d_i)
1877	821	1 (66.5)	21 (53.3)	19 (52.4)	2 (79.4)
1878	769	2 (66.4)	23 (52.9)	21 (51.9)	1 (79.2)
1873	821	3 (66.3)	22 (52.0)	20 (50.9)	4 (79.1)
1870	821	4 (66.3)	19 (46.7)	18 (47.9)	3 (79.1)
1771	779	5 (63.1)	24 (44.8)	26 (47.9)	5 (74.8)
1855	821	6 (63.1)	2 (44.7)	24 (46.6)	6 (74.5)
1854	769	7 (61.2)	30 (43.6)	30 (46.0)	7 (74.2)
1783	220	8 (46.0)	26 (43.3)	25 (45.7)	19 (65.7)
1871	340	9 (43.0)	18 (42.9)	16 (45.4)	22 (59.6)
1826	1524	10 (41.6)	17 (41.5)	15 (44.4)	36 (53.8)
Iteration # 1					
η_1 (rad/s)^2		0.218	1.22	0.22	0.01
P_{C1}		23,811	25,351	25,351	26,203
		(23,973)	(25,871)	(25,833)	(26,210)
Nr of CMs		32	36	36	38

among the 4, while for contingencies Nr 10 and 11 these CMs are located rather far away from the top.

The simulations start with a total power decrease $\Delta P_C = 0.03 * 27,014 = 811$ MW. (Actually, this is the maximum power decrease, imposed by the severest contingency, Nr 30.) Decreasing the initial powers P_{C0} of the 4 contingencies by this ΔP_C yields the P_{C1} values indicated in row 2 of iteration Nr 1. The corresponding new margins are all positive, though for contingency Nr 30 it is almost zero. The P_C values listed between brackets are obtained by interpolating η_0 and η_1.[37]

---

[37] For example, it was found that to stabilize contingency Nr 30, the generation of the 38 machines should be shifted by $27,014 - 26,210 = 804$ MW.

Discussion. Computationally, stabilizing simultaneously all harmful contingencies is more expensive than stabilizing a single one, but less expensive than stabilizing them sequentially.

But the main advantage of the simultaneous stabilization is that it furnishes near optimal solutions.

Of course, stabilizing the most constraining contingency requires overstabilizing the others. For example, in the above illustration, contingencies Nr 1, 10 and 11 are slightly overstabilized: according to their P_C values indicated between brackets in Table 4.16, this overstabilization corresponds to respectively 154, 513 and 475 MW.[38] These powers are, however, negligible in terms of percentage of the total CMs' generation (the maximum decrease is 2 %).

And again, the generation decrease in CMs has been compensated by an (almost) equal increase in NMs without any load shedding.

4. SUMMARY

This chapter has collected all necessary ingredients for performing real-time preventive TSA&C.

Section 2 described various types of stability limit computations: CCTs and PLs; accurate as well as approximate ones.

Section 3 proposed approximate CCT computation techniques able to devise an integrated software called FILTRA, which filters out uninteresting contingencies, ranks potentially interesting ones and finally assesses the harmful contingencies.

Section 4 addressed the important issue of control. Techniques have thus been devised, able to stabilize harmful contingencies, relying on generation power shifting from CMs to NMs. Many patterns were suggested to reschedule generation of CMs taking into account the type of instability, the number of CMs and other important factors. In short, Section 4 has shown that stabilization, i.e., control which has long been considered a problematic issue becomes a straightforward task for SIME. Besides, it has shown that this stabilization, if well designed, is not as "expensive" as used to be considered.

Finally, it was suggested that rescheduling generation of NMs may open avenues to other objectives, additional to meeting transient stability constraints. Chapter 5 will focus on such objectives.

---

[38] For example, for stabilizing contingency Nr 1 alone, one should shift $24,623 - 23,973 = 650$ MW, i.e. $804 - 650 = 154$ MW less.

Chapter 5

INTEGRATED TSA&C SOFTWARE

In this chapter, techniques developed in Chapter 4 are organized to set up a unified TSA&C software aiming to cover an as large as possible spectrum of preventive security issues.

Section 1 concentrates on this unified software, first considered alone, then interfaced with an OPF algorithm, in order to broaden further its possibilities.

In Section 2 this software is applied to a real-world problem: the computation of transient stability-constrained maximum power transfer between areas. Two cases are considered. The one deals with local modes of transient stability constraints; the other with inter-area modes. The use of OPF is shown to be beneficial in many respects: quality of the obtained result (amount of allowable power transfer); computational performances; simultaneous coverage of other types of limits, like static ones (bus voltage and thermal line limits).

Finally, Section 3 envisages various ways of integrating the TSA&C package in an EMS environment in order to perform on-line congestion management and available transfer capability calculations. It also proposes use of this package in a dispatcher training simulator environment.

The various developments illustrate the flexibility of the integrated TSA&C software, and its ability to comply with any type of preventive studies, be it for planning, operational planning or real-time operation.

1. INTEGRATED SOFTWARE
1.1 Basic TSA&C software

The functions developed in Chapter 4 to conduct a complete TSA&C task are combined here to get an integrated TSA&C package. Subsequently, this basic software will be augmented with an OPF program so as to meet transient

140   TRANSIENT STABILITY OF POWER SYSTEMS

stability constraints together with other constraints imposed in EMS environments.

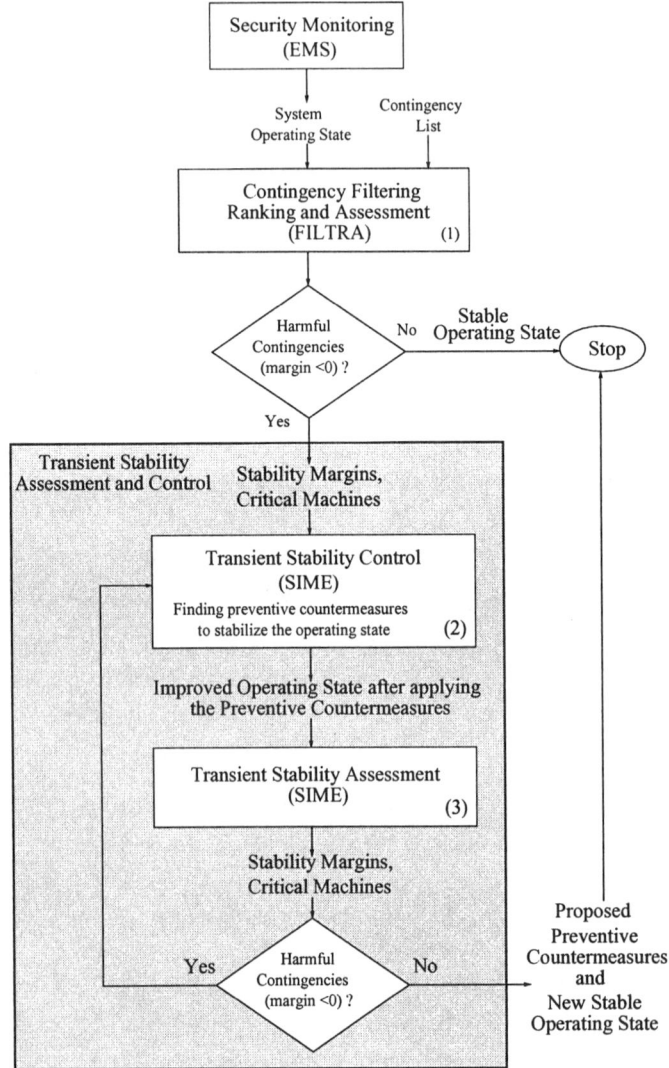

Figure 5.1.   Organization of the preventive TSA&C software package

Figure 5.1 displays the basic integrated software package, that we shortly describe below.

(i) The output of the state estimator furnishes the data for running a power flow to determine the current operating state. This, together with the contingency set are injected to FILTRA.

(ii) FILTRA (Block (1) in Fig. 5.1) identifies the harmful contingencies from this generally very large initial list of plausible contingencies. Further, for each one of the harmful contingencies, it provides the stability margin and corresponding CMs.

(iii) Upon receipt of this information, the Transient Stability Control Block (Block (2) in Fig. 5.1):

- determines the corresponding control actions (active power change in each CM) for each one of the harmful contingencies;
- combines the resulting control actions to compute the amount of active power change in each CM necessary to stabilize the set of harmful contingencies simultaneously;
- reallocates the corresponding amount of active power on NMs so as to encounter predefined objective functions, additional to the transient stability constraints; let us recall that this reallocation on NMs affects only marginally - if at all - transient stability.

(iv) The Transient Stability Assessment Block (Block (3) in Fig. 5.1) receives the "improved" operating state found by the power flow program after applying the preventive control actions (in this case, generation rescheduling), and assesses power system transient stability with respect to the previous set of harmful contingencies. If the assessment block finds that some contingencies are still unstable, it computes the corresponding margins, identifies the CMs and sends this information back to the Transient Stability Control block. If the power system is stable for all contingencies, the new operating state is declared to be stable and the process stops.

The cycle: "identification of the harmful contingencies, their corresponding margins and CMs - computation of active power reallocation on each one of them - reallocation of the corresponding power among NMs" is repeated until stabilizing all harmful contingencies.

Generally, two to three iterations are sufficient.

A final check is then performed to verify that the procedure has not destabilized any of the previously potentially harmful contingencies (mandatory) and harmless contingencies (optional).

1.2 Multi-objective TSA&C software

The above important achievement in real-time TSA&C may further be broadened by interfacing the TSA&C block of Fig. 5.1 with an OPF software. Indeed, this "augmented" software package allows extending TSA&C along at least the following two directions (Fig. 5.2).

(i) At the input of Block (1), by determining an operating state which meets a predefined objective function (e.g., maximum allowable transfer on a cut-

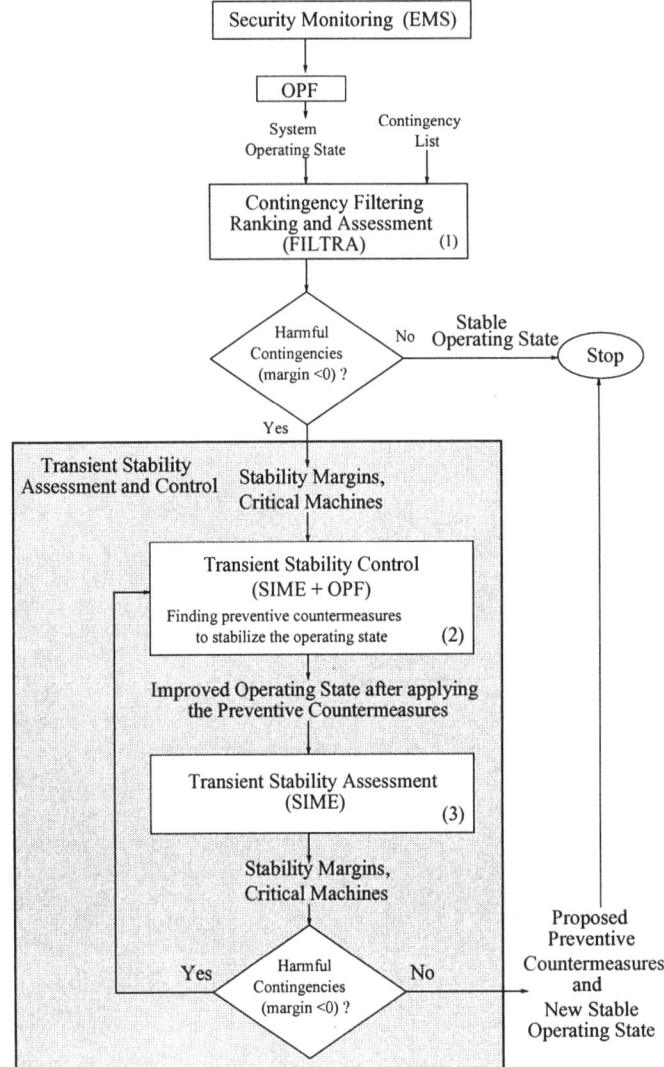

Figure 5.2. Organization of the TSA&C-OPF software package

set, congestion management, etc.), while satisfying static constraints of the transmission network (bus voltage and power line limits).

(ii) At the output of Block (2), by reallocating active power on NMs so as to meet the predefined objective while satisfying above static constraints.

At the same time, the combined use of OPF contributes to the overall integration of all security functions in the EMS environment, discussed below, in Section 3.

The particular way the OPF software is adapted to cope with TSA&C software requirements is briefly described below.

1.3 Adapting the basic OPF algorithm

Figure 5.1 shows that transient stability constraints are satisfied by rescheduling the generation of CMs. To meet other types of objectives, OPF acts on NMs. The resulting integrated TSA&C and OPF software is portrayed in Fig. 5.2.

As an example, the formulation of the OPF problem to maximize the power flow in the set of interface lines of interest, is described by the following equations:

$$\text{Max} \left\{ IF_t = \sum_{il \in nli} IF_{il} \right\} \quad (5.1)$$

s.t.

$$P_k = \sum_{j \in k} V_k V_j \left[G_{kj} \cos(\delta_k - \delta_j) + B_{kj} \sin(\delta_k - \delta_j) \right] , \quad k \in N \quad (5.2)$$

$$Q_k = \sum_{j \in k} V_k V_j \left[G_{kj} \sin(\delta_k - \delta_j) - B_{kj} \cos(\delta_k - \delta_j) \right] , \quad k \in N \quad (5.3)$$

$$P_i^m \leq P_i \leq P_i^M \quad , \quad i \in G \quad (5.4)$$
$$Q_i^m \leq Q_i \leq Q_i^M \quad , \quad i \in G \quad (5.5)$$
$$V_k^m \leq V_k \leq V_k^M \quad , \quad k \in N \quad (5.6)$$
$$|S_l| \leq S_l^M \quad , \quad l \in T \quad (5.7)$$

where the following notation is used:

IF_t	total interface flow
IF_{il}	interface flow through interconnecting line il
P_k	active power injection in bus k
Q_k	reactive power injection in bus k
V_k	voltage magnitude in bus k
δ_k	voltage angle of bus k
S_l	apparent power through line l
G_{kj} and B_{kj}	elements of the bus admittance matrix.

G is the set of all generator busses, N the set of all buses. T denotes the set of system's lines and nli the set of the interconnecting lines of interest. m and M denote lower and upper limits respectively.

Equation (5.1) represents the objective function, eqs (5.2) and (5.3) are the network equations, and (5.4) and (5.5) are generation limits. Static security constraints are represented by eqs (5.6) and (5.7) [Bettiol et al., 1999b].

After receiving margins and CMs from FILTRA (Block (1) of Fig. 5.2), the Transient Stability Control block (Block (2) of Fig. 5.2) provides the new values of the upper limit (or the lower limit, depending on the instability phenomena) for the active power output of each CM necessary to meet the transient stability constraints. Note that the power generation of these CMs is not going to be fixed at these values but only bounded by them, and that their final values are going to be found after the optimization process. In the case of NMs, their limits do not change and their power outputs depend only on the objective function.

Finally, the OPF study gives a new optimal steady state operating state that takes into account not only transient stability and static constraints, but also maximizes the power flow in interface lines. This new operating state is used by SIME for assessing all harmful contingencies. If the system is stable for the whole set of contingencies, the process stops; if not, new preventive control actions are computed and another OPF is run.

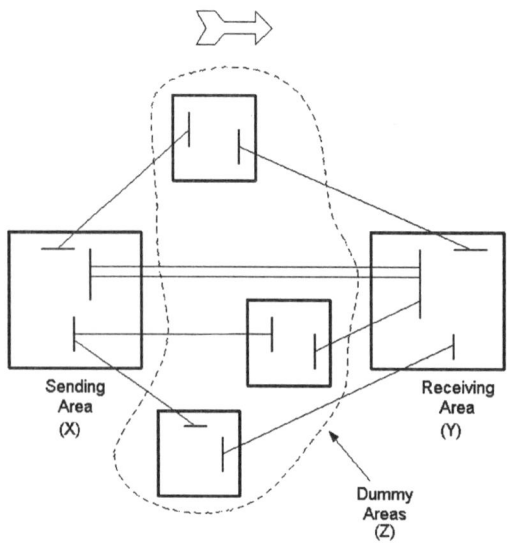

Figure 5.3. Power transfer among areas of an interconnected power system

2. A CASE-STUDY

2.1 Maximum allowable transfer: problem statement

Consider an interconnected power system operating with inter-area power transfers as schematically shown in Fig. 5.3, where three different areas are distinguished: the exporting area (denoted X), the importing area (Y) and the set of "non participating" areas (Z), i.e., areas which are not directly involved in the power transfer between areas X and Y. Assume that the amount of power to be transferred between areas X and Y is transient stability-limited, i.e., that for this amount there exists (a set of) harmful contingencies. Under these conditions, the Maximum Allowable Transfer (MAT) problem addresses the following question: "determine the transient stability-constrained maximum power transfer from area X to area Y ".

This problem may be decomposed into the following sub-problems:

(i) determine the initial operating (base-case) state which achieves a maximum power transfer X → Y;
(ii) apply to this base-case the set of plausible contingencies, and determine whether there are some constraining, i.e., harmful ones;
(iii) if yes, apply procedures of § 4.3 of Chapter 4 to assess the necessary generation power decrease in CMs[1] and its distribution among them;
(iv) at the same time, report the corresponding (almost equal amount of) power increase in NMs in such a way so as to achieve maximum power transfer.

Obviously, both schemes of Figs 5.1 and 5.2 can properly handle sub-problems (ii) and (iii). On the other hand, sub-problems (i) and (iv) can be handled by using either proper logical rules, or the OPF algorithm of Fig. 5.2.

Both approaches have been investigated on the Brazilian South-Southeast power system described in Section 4 of Appendix B [Bettiol et al., 1999a]. An in-depth description of these investigations is given in [Bettiol, 1999], where two types of transient stability-limited explorations have been carried out: one concerns plant mode instabilities, the other inter-area mode instabilities. Main results of this large-scale investigation are reported below, in §§ 2.2, 2.3.

2.2 Plant mode instability constraints

2.2.1 Problem description

The objective here is to determine the transient stability-limited maximum power transfer in the South part of the system (see Fig. 5.4) between areas 12 to 15 and area 16 which operates under a generation deficit. This transfer is limited by a "local" or "plant" mode, as will appear below.

To search this "local" power transfer limit, only the 17 machines of the South power system are included in the generation rescheduling procedure. The active power dispatch of the remaining machines of the Southeast-Centerwest power system (about 27,445 MW) is kept unchanged during the search. All machines of the exporting areas operate in their upper generation power limits (6,394 MW) and the machines of the importing area in their lower generation limits (918 MW). Thus, the computed base case corresponds to a large amount of power (about 2,621 MW) transferred from areas 12 to 15 of the South power

---
[1] Remember, generation decrease in CMs corresponds to upswing instabilities, while generation increase to backswing ones.

146 TRANSIENT STABILITY OF POWER SYSTEMS

system to area 16. Figure 5.4 portrays schematically the tie-lines of interest for the desired operating condition.

To check whether the maximum power transfer limit established in steady-state meets also transient stability constraints, the MAT problem is subdivided into the following tasks.

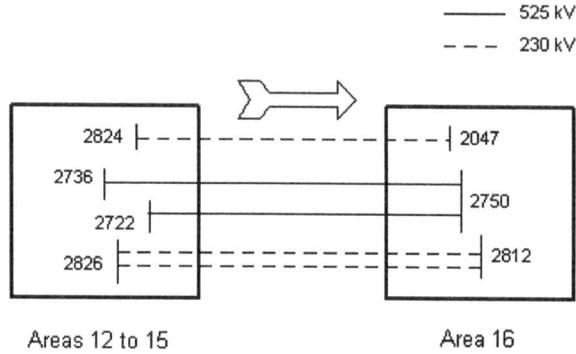

Figure 5.4. Schematic representation of the tie-lines of interest. Adapted from [Bettiol, 1999]

2.2.2 Contingency filtering, ranking, assessment

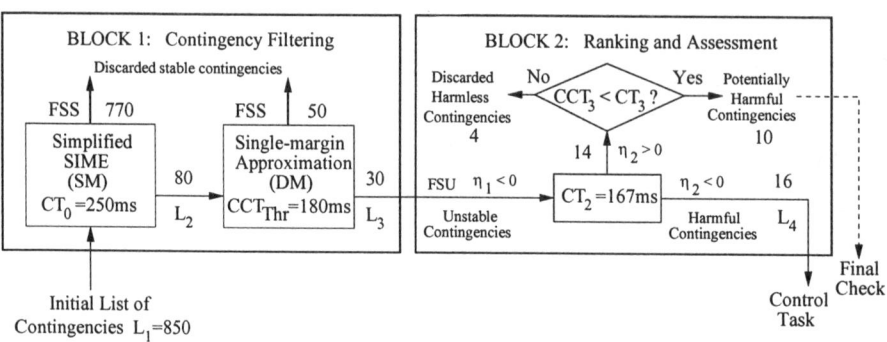

Figure 5.5. A realization of FILTRA with a special purpose filtering block. First case study.

Block 1: Contingency filtering (Fig. 5.5). The initial list of plausible contingencies (list L_1) comprises 850 contingencies. They are processed according to the general two-block organization of FILTRA displayed in Fig. 4.8 where, however, Block 1 is composed of two sub-blocks. (See Fig. 5.5).

Sub-block Nr 1 is made of a special-purpose SIME program which uses fixed large size T-D integration steps (e.g., of 25 ms), a short maximum integration period (1 s) and a $CT_0 = 250$ ms. The power system is modeled in the conventional simplified way (constant emf behind transient reactance; constant mechanical power; constant impedance loads).

Thus, with $CT_0 = 250$ ms, and using the first-swing stopping criteria of § 2.4.4 of Chapter 4, the program declares a contingency either FSS and discards it, or FSU and includes it in list L_2.

Sub-block Nr 2 uses detailed power system model (DM). It relies on the compensation scheme (c) of Fig. 3.11, which provides an approximate first-swing CCT, labeled t_{c3}, for all contingencies found to be unstable (i.e., having a negative margin, η_1). Thus, for a clearing time, CT_1, a contingency

- found to be FSS or to have a $t_{c3} > 180$ ms, is discarded
- having a $t_{c3} < 180$ ms, is stored in list L_3 and sent to the ranking-assessment block.

Note that, here, CT_1 is not constant, but determined according to the initial clearing time conditions of § 2.2.3 of Chapter 4. This CT_1 along with the corresponding margin, η_1, is sent to Block 2 for contingency ranking and assessment.[2]

Block 2: Contingency ranking and assessment (Fig. 5.5). Following the general pattern, FILTRA considers the FSU contingencies and their (positive or negative) margin, η_2, using detailed power system modeling (DM), and $CT_2 = 167$ ms.

In this case study, out of the 30 FSU contingencies:
16 are found to be harmful $(\eta_2 < 0)$
10 are found to be potentially harmful $(CCT_3 < CT_3 = 180 \text{ ms})$
4 are found to be harmless $(CCT_3 > 180 \text{ ms})$.

- The harmful contingencies are stabilized below, in § 2.2.3.

  Table 5.1 gathers FILTRA's information about them:

  Columns 2, 3 list the first clearing time, CT_1, and the corresponding normalized margin, η_1

  Column 4 lists the second margin, η_2, corresponding to the second (fixed) clearing time, $CT_2 = 167$ ms

  Column 5 gives the contingencies CCTs computed by extrapolating linearly

---

[2] Let us stress that design of Block 1 of Fig. 5.5 follows the general pattern of Fig. 4.8 but is different from that of Fig. 4.9 in various respects. In particular, by the use of a simplified power system model (sub-block 1) and of a variable CT_1 (sub-block 2). Besides, Block 2 uses $CT_2 = 167$ ms, i.e. larger than that of Fig. 4.9.

Table 5.1. Harmful contingencies ranking and assessment by FILTRA

1	2	3	4	5	6
Cont. Nr	CT_1 (ms)	η_1 (CT_1)	η_2 ($CT_2 = 167$ ms)	CCT (ms)	Ranking
826	240	-46.69	-11.81	–	1
827	190	-36.46	-23.48	125	2
784	190	-29.98	-18.59	129	3
831	230	-49.21	-11.48	148	4
802	210	-46.14	-13.85	148	5
830	230	-49.68	-8.76	153	6
841	250	-73.43	-8.40	156	7
816	250	-73.42	-8.58	156	8
832	230	-48.42	-6.93	157	9
744	250	-72.84	-7.72	157	10
843	250	-72.8	-7.58	157	11
828	240	-49.12	-5.14	159	12
715	250	-72.83	-6.68	160	13
842	250	-72.83	-5.33	160	14
704	220	-103.46	-7.38	163	15
701	220	-100.78	-0.41	167	16

the above two margins.[3] This provides another way of ranking contingencies, more reliable than, though as unexpensive as t_u used in Chapter 4.

- The potentially harmful contingencies will be stored for a stability check under the new operating conditions resulting from the stabilization of the harmful contingencies.

- The harmless contingencies are discarded. They might be reconsidered for a final check of all 850 contingencies, if deemed necessary.

Summary of contingency screening. Figure 5.5 summarizes the FILTRA results. It shows that, out of the 850 screened contingencies (list L1), the first sub-block discards 770 (90.6 %) of them, and sends 80 to the second sub-block (list L2). Similarly, the second sub-block eliminates 50 contingencies (5.9 %), "sending" 30 (3.5 %) to Block 2; these 30 contingencies compose list L3 (FSU contingencies). Finally, FILTRA identifies 16 harmful, 10 potentially harmful and 4 harmless contingencies.

---

[3] in all cases, except for the first contingency (Nr 826) where the CMs are not the same for the two stimulations.

Chapter 5 - INTEGRATED TSA&C SOFTWARE

To summarize, FILTRA finds that the generation pattern and corresponding power transfer level between exporting and importing areas of the base case conditions are actually constrained by 16 harmful contingencies.

2.2.3 Contingencies' simultaneous stabilization

Base case. As stated earlier, the base case aims at achieving the "maximum" power transfer between areas X and Y. Actually, this amount will be different when determined heuristically or via an OPF program.

However, here, to allow easy comparison of the two stabilization procedures, we consider the common base case used so far by FILTRA, obtained by an OPF program. From this base case, the iterations for stabilizing the contingencies will be carried out on one hand by the logical rules defined below, on the other hand by an OPF program [Granville, 1994, Mello et al., 1997].

Table 5.2 gathers the resulting solutions that we comment hereafter.

Column 3: time to instability t_u (maximum integration period, MIP). This time expresses the computational requirements in terms of sTDI (actually, in msTDI)

Column 4: type of the OMIB swing: +1 means single upswing[4]

Column 5: total amount of generation decrease necessary to cancel out the unstable margin of Column 2, as provided by the compensation scheme of § 3.3 of Chapter 3

Column 6: identification of the CM(s) classified in decreasing order of criticalness, assessed in terms of the angular distance (d_i) of each CM_i.

Recall that to stabilize all harmful contingencies simultaneously, the amount of generation power decrease in each CM is chosen to be the largest among those listed in Column 6 for all contingencies that contain this CM (cfr § 4.3 of Chapter 4). In other words, the procedure consists of monitoring only the contingencies which contain the most severely disturbed CMs. Hence, only contingencies 827, 784, 802 and 704 are monitored.

The resulting generation decrease (i.e, the sum of power excess in machines 2707 (48 MW), 2706 (30 MW), 2705 (15 MW), 2704 (6 MW) and 2573 (5 MW), 2712 (76 MW)) totals 180 MW.

Below we reallocate these 180 MW on NMs according to (I) logical rules, and (II) OPF.

I. NMs' generation rescheduling via logical rules. Basically, the following three selection rules are used

---

[4] Obviously, in Table 5.2 all contingencies are of type 1; but in Table 5.9, § 2.3 there are of various other types

150   TRANSIENT STABILITY OF POWER SYSTEMS

#1: available generation, complying with lower and upper machines' power limits

#2: generation rescheduling for maximizing the power transfer between areas. For example, it is obvious that the power transfer will increase by increasing the generation in a NM if this NM belongs to area X

#3: additional criteria; for example generation cost.

These criteria, gathered in a "logical table" [Bettiol et al., 1999a, Bettiol, 1999], lead to the stabilization procedure summarized in Table 5.2 and commented below.

Base case. Selection rule #1 compares the current values of power generation of the NMs with their operational upper limits[5] so as to reschedule generation of only those NMs that operate below their upper generation limits.

Selection rule #2 advocates considering the NMs of exporting areas. However, all these NMs operate in their upper generation limit. One is therefore led to increase generation in the 8 NMs of the importing area 16, by considering their available generation power. Thus, the amount of 180 MW is distributed among these 8 machines proportionally to their maximum power generation capacities.

A new power flow is then performed with the indicated generation rescheduling of both CMs and NMs. Using the resulting new operating state, SIME is run again to assess the 4 more constraining harmful contingencies under monitoring.[6]

First iteration. Here, only one contingency has still negative margin imposing a further generation decrease in machine 2573. Note that the positive margins suggest also minor readjustments.

Second iteration. After a new power flow accounting for these small adjustments, SIME indicates that contingency 704 is now stabilized: the iterative procedure is thus finished.

Final check. Concerning a final check, let us observe again the following.

1. In real-time preventive TSA&C (which could be performed, for instance, each 15 or 20 minutes), a final check could be mandatory only for the harmful and potentially harmful contingencies.

2. For the remaining contingencies (i.e., those initially discarded by FILTRA), the final check could be optional; for example, it could be performed after significant modifications of the actual operating state (resulting from significantly different generation patterns, load scenarios, topology changes,

---

[5] $P_{\max}$ reported in [Bettiol, 1999].
[6] Remember, stabilizing contingencies 827, 784, 802 and 704 stabilizes all other contingencies as well.

Table 5.2. Simultaneous stabilization of all harmful contingencies via: (i) logical rules; (ii) OPF. Adapted from [Bettiol, 1999]

1	2	3	4	5	6
				Base case	
Cont. Nr	η_2	t_u (MIP) (ms)	Swing	ΔP_C (MW)	CM_i ($d_i(°)$, ΔP_{Ci}(MW))
826	-11.81	355	+1	-38	2707(36,-22), 2706(31,-16)
827	-23.48	305	+1	-78	2707(45,**-48**), 2706(33,**-30**)
784	-18.59	315	+1	-90	2707(102,-39), 2706(90,-30), 2705(88,**-15**), 2704(52,**-6**)
831	-13.85	375	+1	-52	2707(118,-25), 2705(89,-8), 2706(88,-16), 2704(53,-3)
802	-13.85	375	+1	-76	2712(91,**-76**)
830	-8.76	390	+1	-58	2706(115,-21), 2707(103,-23), 2705(100,-10), 2704(62,-4)
841	-8.40	410	+1	-44	2712(94,44)
816	-8.58	415	+1	-44	2712(95,-44)
832	-6.93	405	+1	-29	2705(106,-5), 2707(106,-12), 2706(104,-10), 2704(69,-2)
744	-7.72	415	+1	-39	2712(95,-39)
843	-7.58	420	+1	-38	2712(96,-38)
828	-5.14	420	+1	-19	2705(114,-4), 2706(102,-6), 2707(98,-7), 2704(70,-2)
715	-6.68	420	+1	-34	2712(93,-34)
842	-5.33	445	+1	-27	2715(95,-27)
704	-7.38	370	+1	-5	2573(105,**-5**)
701	-0.41	465	+1	-1	2573(103,-1)
				Logical rules: first iteration	
827	1.08	(4,000)	–	+3	2707(–,+2), 2706(–,+1)
784	2.60	(4,000)	–	+12	2707(–,+5), 2706(–,+4), 2705(–,+2), 2704(–,+1)
802	0.07	(4,000)	–	0	2712(–,0)
704	-1.34	440	+1	-2	2573(102,**-2**)
				Logical rules: second iteration	
704	0.50	(4,000)	–	0	2573(–,0)
				OPF: one sole iteration	
827	1.51	(4,000)	–	+5	2707(–,+3), 2706(–,+2)
784	3.27	(4,000)	–	+15	2707(–,+6), 2706(–,+5), 2705(–,+3), 2704(–,+1)
802	1.33	(4,000)	–	+7	2712(–,+7)
704	5.21	(4,000)	–	+2	2573(–,+2)

maintenance schedules, etc.), or upon request of the system operator. This final check could be performed in off-line mode.

152   TRANSIENT STABILITY OF POWER SYSTEMS

II. NMs' generation rescheduling via OPF. In this approach, the OPF program reallocates automatically the amount of 180 MW in NMs, with the objective to maximize the power transfer on the tie-lines of Fig. 5.4; at the same time, this program determines the new operating state. SIME is then run again to assess the existence and severity of harmful contingencies.

Last part of Table 5.2 summarizes this stability assessment and shows that all harmful contingencies are stabilized after one iteration.

2.2.4 Logical rule vs OPF-based procedures

Below, we compare the results of the two approaches described above with respect to:

- generation power rescheduling of CMs and NMs
- resulting power exchange between exporting and importing areas under transient stability constraints
- resulting total losses
- reliability
- computational requirements.

Generation rescheduling of CMs and NMs.   Table 5.3 compares the rescheduling patterns. In Column 1, the asterisk identifies the CMs. In Columns 3 and 4, the bold-face values between brackets indicate the generation power changes imposed respectively by logical rules and OPF.

Observe that the differences of the solutions provided by logical rules and by OPF are essentially marked on NMs, while they are quite marginal for CMs, which goes along the very strategy of SIME-based stabilization procedures.

Incidentally, note that the OPF procedure imposes on the slack bus to furnish its maximum generation capacity; more generally, its generation rescheduling pattern is quite different from that of the logical rules. This is a direct consequence of the "optimal" generation rescheduling of the NMs provided by OPF.

Interface power flows.   Table 5.4 summarizes the changes in the interface flows imposed by the two stabilization procedures. Columns 2 to 5 list the power flows (measured on the bus of the exporting area) on the tie-lines of interest. The total power flow leaving the exporting area is shown in Column 6.

Observe that the total interface flow decrease imposed by the logical rules and by OPF is respectively 208 MW (7.94 %) and 151 MW (5.76 %). Obviously, this latter is closer to the maximal, steady-state unconstrained initial value obtained by the OPF program. This is a direct consequence of the better generation rescheduling of NMs determined by OPF.

Figure 5.6 displays the differences resulting from the two approaches.

Table 5.3. Generation rescheduling patterns: logical rules vs OPF. Adapted from [Bettiol, 1999].

1	2	3		4	
	Generation power (MW)				
Machine	Initial	Final			
	(Base-case)	Logical rules		OPF	
2094	300	337	(+37)	303	(+3)
2101	90	107	(+17)	94	(+4)
2174	60	72	(+12)	63	(+3)
2176	120	157	(+37)	123	(+3)
2183	80	91	(+11)	84	(+4)
2569	1,672	1,672		1,672	
2573*	248	241	(-7)	243	(-5)
2674 (slack)	1,260	1,215	(-45)	1,260	
2702	30	35	(+5)	33	(+3)
2704*	100	94	(-6)	94	(-6)
2705*	132	117	(-15)	117	(-15)
2706*	250	220	(-30)	220	(-30)
2707*	350	302	(-48)	302	(-48)
2710	80	106	(+26)	190	(+110)
2712*	1,050	974	(-76)	974	(-76)
2714	1,332	1,332		1,332	
2769	158	193	(+35)	161	(+3)

Table 5.4. Interface flow changes: logical rules vs OPF. Adapted from [Bettiol, 1999]

1	2	3	4	5	6
	Interface flows (MW)				
Case	525 kV		230 kV		
	2736-2750	2722-2750	2826-2812	2824-2047	**Total**
Initial	1,141.5	949.8	308.0	221.7	**2,621**
Final Logical rules	1,073.9	897.0	259.7	182.4	**2,413**
OPF	1,110.5	927.2	246.4	186.0	**2,470**

Active power losses. Table 5.5 shows the changes in active power losses for the entire power system (exporting, importing, and "remaining" areas), resulting from the generation rescheduling. Note that the OPF allows a larger reduction of the total active power losses: 46 MW (2.51 %) for OPF vs 2 MW (0.11 %) for the logical rules. This might be attributed to the fact that the static constraints (especially bus voltages and power limits) are met by the OPF approach at each step of the iterative procedure.

154   TRANSIENT STABILITY OF POWER SYSTEMS

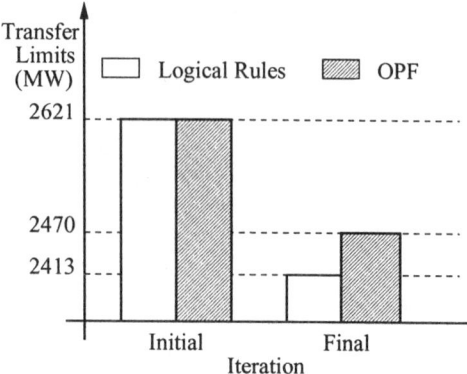

Figure 5.6. Maximum transient-stability constrained power transfer limits. Adapted from [Bettiol, 1999]

Table 5.5. Active power losses: logical rules vs OPF. Adapted from [Bettiol, 1999]

1	2	3	4	5	6	7	8	9
	\multicolumn{8}{c}{Active power losses (MW)}							
Iteration	\multicolumn{4}{c}{Logical rules}			\multicolumn{4}{c}{OPF}			
	Areas 01 to 11	Areas 12 to 15	Area 16	Total	Areas 01 to 11	Areas 12 to 15	Area 16	Total
Initial	1,373	376	86	**1,835**	1,373	376	86	**1,835**
Final	1,385	371	77	**1,833**	1,367	344	78	**1,789**

Reliability. The reliability of the two procedures was checked by comparing the resulting CCTs with those obtained by the ST-600 program (with simplified and detailed power system modeling, as appropriate), used here as the reference.[7]

For the proposed contingency filtering and assessment task, it was found that all discarded contingencies are, indeed, harmless, i.e. that their CCTs are larger than the threshold values at each step of the contingency filtering, assessment and final check.

Similarly, as a validation of the overall procedure, the CCTs of all 850 contingencies have been again computed by the ST-600 program at the end of the iterative procedure. Table 5.6 displays the initial and final CCT values of the previously 16 harmful contingencies (in bold) and 10 potentially harmful ones. Note that:

---

[7]In these case-studies SIME is coupled with the Hydro-Québec T-D stability program, ST -600.

Table 5.6. Changes in (potentially) harmful contingencies' CCTs

1	2	3	4
		CCT values (ms)	
Contingency Nr	Initial	Final	
		Logical rules	OPF
827	115	173	173
784	120	178	178
826	133	197	197
831	134	188	192
802	143	168	168
830	144	197	197
828	147	198	198
832	148	197	198
744	152	173	178
816	152	173	178
841	152	173	178
843	152	173	178
715	157	178	178
842	159	178	183
704	161	168	168
701	165	173	173
824	168	222	222
796	169	183	173
825	169	212	197
834	169	216	222
103	170	168	168
726	171	178	178
140	173	181	172
809	177	199	193
112	178	168	169
725	180	183	188

- after the generation rescheduling by the two approaches, the CCT values of the previously harmful contingencies (marked in bold in Table 5.6), are quite similar, and close to the adopted threshold (167 ms)
- the CCT values of contingencies 103, 112, and 140 are smaller after the generation rescheduling (this latter with the OPF approach) but never smaller than 167 ms.

Computational requirements. Tables 5.7 and 5.8 list the computational requirements of the various tasks of the two MAT procedures. The simulations were run on a single SUN UltraSPARC-II workstation (Ultra-10 model, 300 MHz, 1024 MB of RAM).

156   *TRANSIENT STABILITY OF POWER SYSTEMS*

Table 5.7. Computational requirements of the procedure using logical rules

Task Nr	Task identification	Computational requirements		
		sTDI	seconds	%
1	**OPF (base case)**	**0.20**	**5**	**0.11**
2	FILTRA			
	Filtering block			
	First sub-block (SM)	5.00	125	2.86
	Second sub-block (DM)	30.53	804	18.42
	Assessment block	62.28	1,672	38.29
	Total FILTRA	**97.81**	**2,601**	**59.57**
3	**Stabilization**	**16.44**	**421**	**9.14**
4	Final check			
	Harmful and potentially harmful ctgs (26)	44.29	1,163	26.62
	Remaining contingencies (824)	7.00	177	4.06
	Total final check	**51.29**	**1,340**	**30.68**
	TOTAL	**165.74**	**4,367**	**100.00**

Globally, they correspond to about 73 minutes (using the logical rules) and 77 minutes (using the OPF program), respectively.

Table 5.8. Computational requirements of the procedure using OPF

Task Nr	Task identification	Computational requirements		
		sTDI	seconds	%
1	**OPF (base case)**	**0.20**	**5**	**0.11**
2	FILTRA			
	Filtering block			
	First sub-block (SM)	5.00	125	2.71
	Second sub-block (DM)	30.53	804	17.46
	Assessment block	62.28	1,672	36.30
	Total FILTRA	**97.81**	**2,601**	**56.47**
3	**Stabilization**	**16.16**	**387**	**8.40**
4	Final check			
	Harmful and potentially harmful ctgs (26)	56.73	1,452	31.53
	Remaining contingencies (824)	6.40	161	3.49
	Total final check	**63.13**	**1,613**	**35.02**
	TOTAL	**177.30**	**4,606**	**100.00**

A more detailed inspection of the above tables suggests that

- usage of OPF affects rather marginally the computing effort
- stabilization requires less than 10 % of the overall computing effort
- FILTRA, together with the final check, are the most time consuming procedures. Note however that their computations can very easily be distributed among many computers.

To fix ideas, using 5 PCs would reduce the overall computing effort to about 20 min, in both cases (logical rules or OPF). This CPU could further be reduced if the final check was restricted to a small sub-set of contingencies, see in § 2.2.3. Additional speeding up procedures could easily be thought of.

2.3 Inter-area mode instability constraints

2.3.1 Problem description

While in the previous case the power was transferred between areas of the same region (Brazilian South system), this case considers the power transferred between different regions (Brazilian Southeast-Centerwest and South systems), where each region is composed of several areas.

This breeds various differences with the previous study. In particular:

- inter-area mode instabilities, resulting in multi- and back-swing phenomena rather than local mode instabilities;
- larger size of the optimization problem, which involves 56 machines instead of 17;
- use of logical rules becomes more problematic.

Thus, in what follows, the transient stability-constrained MAT problem will be solved by using OPF only. The comparative study with the logical rules-based procedure may be found in [Bettiol, 1999].

2.3.2 Base case conditions

The OPF program determines the steady-state maximum power transfer limit between exporting and importing areas (i.e., the base case conditions). This base case is set up with the same objective function, control variables, and functional constraints as in § 2.2.

Here, the MAT problem concerns the transient stability-constrained maximum power transfer between the Southeast-Centerwest (areas 01 to 11) and South (areas 12 to 16) power systems, see Fig. 5.7. For the computed base case, the 39 machines of the exporting area operate with 94 % (i.e., 28,969 MW) of their maximum generation power capacity, while the 17 machines of the importing area operate with their lower power generation limit (i.e., 4,784 MW, which is equivalent to 58 % of the maximum generation capacity of this part of the power system). In this way, the generation deficit in the South power

158   TRANSIENT STABILITY OF POWER SYSTEMS

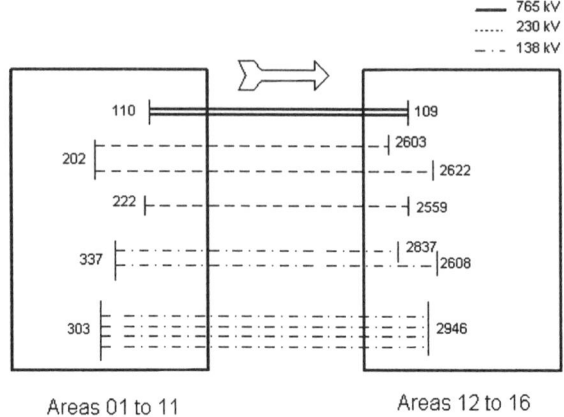

Figure 5.7.   Schematic representation of the tie-lines of concern. Adapted from [Bettiol, 1999]

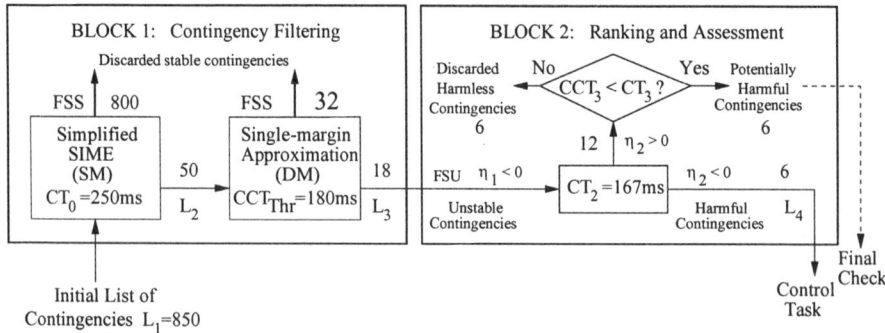

Figure 5.8.   Contingency classification by FILTRA. Second case study.

system is compensated with a large power import (about 3,545 MW) from the Southeast-Centerwest power system. Figure 5.7 shows schematically the tie-lines linking the two power systems under monitoring.

The transient stability-constrained maximum power transfer will be computed following the general organization of Fig. 5.2, whose various sub-tasks are briefly commented below.

2.3.3   Contingency filtering, ranking, assessment

The FILTRA scheme of Fig. 5.5 is again used, with same parameter values. The result is displayed in Fig. 5.8. It shows that the filtering block discards 832 (i.e., 98 % of the) contingencies. Further, from the 18 remaining contingencies, the second block identifies 6 harmful contingencies, whose main features are described below, in Table 5.9. It is interesting to zoom in physical phenomena of some cases reported in this table, before focusing on their stabilization.

2.3.4 Contingencies' simultaneous stabilization

Base case: physical phenomena description. Obviously, the above-identified 6 harmful contingencies suggest the need for generation shifting with respect to the base case.[8]

Let us describe some peculiar physical phenomena caused by some contingencies[9] under this base case power transfer.

A first type of multiswing instability is caused by **contingency Nr 159** (backswing) which causes loss of synchronism because machine 2706 (a medium size thermal power plant) decelerates during its third oscillation, as shown in Fig. 5.9. This contingency, which is located at the 765 kV bulk transmission system of the Itaipu power plant and near the most important transformers linking the Southeast-Centerwest and South power systems, causes the acceleration of all machines of the Southeast-Centerwest power system and of some large machines of the South power system near these interfaces (due to their important synchronizing torques). This causes the deceleration of all thermal power plants (machines 2174, 2176, 2702, 2704, 2706, 2707, and 2769), which are blocked by an under-frequency control device, and of machine 2101 (a small hydro power plant). All these machines are located at the South power system. For this stability scenario, the loss of synchronism main mechanism is the inter-area mode between the Itaipu hydro power station (equivalent machine 16, which is located in the Southeast-Centerwest power system) and thermal power plants of the South power system.

A second type of multiswing instability is caused by **contingency Nr 167**, illustrated in Fig. 5.10. This contingency leads to loss of synchronism by acceleration of machine 21 (located at the Southeast-Centerwest power system) during its second oscillation.

Another inter-area mode restricting the initial power transfer is caused by **contingency Nr 239**, described in Fig. 5.11. This contingency causes 10 machines (located at a same hydropower generation site of the Southeast-Centerwest power system) to accelerate together, causing the system loss of synchronism during their first oscillation.

Stabilization. All these phenomena are properly detected and treated by SIME, as shown in parts (b) of Figs 5.9 to 5.11. The first part of Table 5.9

---

[8] We specify that all these contingencies are located in the Southeast-Centerwest power system (i.e., in the importing area); further, we mention that contingencies 112, 140, and 159 are located at 765 kV buses, while contingencies 239 and 263 at 440 kV buses [Bettiol, 1999].
[9] Recall that the contingencies are cleared after 167 ms.

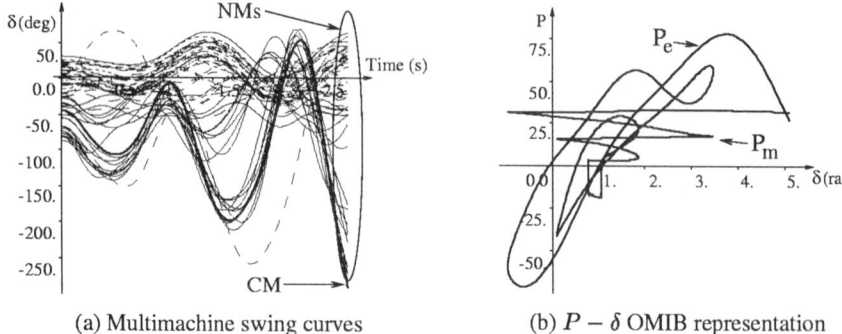

(a) Multimachine swing curves (b) $P - \delta$ OMIB representation

Figure 5.9. Loss of synchronism by backswing multiswing instability phenomena. Contingency 159; $t_e = 167$ ms. Base case operating conditions. Adapted from [Bettiol, 1999]

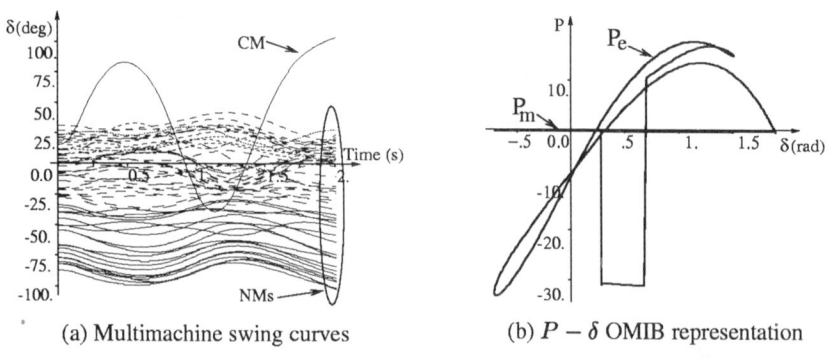

(a) Multimachine swing curves (b) $P - \delta$ OMIB representation

Figure 5.10. Loss of synchronism by multiswing instability phenomena. Contingency 167; $t_e = 167$ ms. Base case operating conditions. Adapted from [Bettiol, 1999]

summarizes their characteristics, including their CCTs found, as previously, by extrapolating linearly margins η_1 and η_2 of FILTRA (see Fig. 5.8).

Concerning the amount ΔP_C, listed in Column 6 of the table, recall that:

- ΔP_C is negative for upswing instabilities and positive for backswing ones;
- the use of the compensation scheme (§ 3.3 of Chapter 3) is not applicable to multiswing instabilities. For such cases (here for Contingencies Nr 159 and 167) ΔP_C is taken arbitrarily 10 % of the actual generation of the corresponding CMs.

Accordingly, the individual generation power decrease (for upward oscillations) or increase (for backward oscillations) for the 13 CMs, are indicated in bold in Table 5.9. Hence, the total generation power decrease required for the base case is 911 MW, i.e., the sum of the following CMs' generation shifts: 16 (-252 MW), 21 (-139 MW), 179 (-63 MW), 181 (-115 MW), 182 (-46 MW),

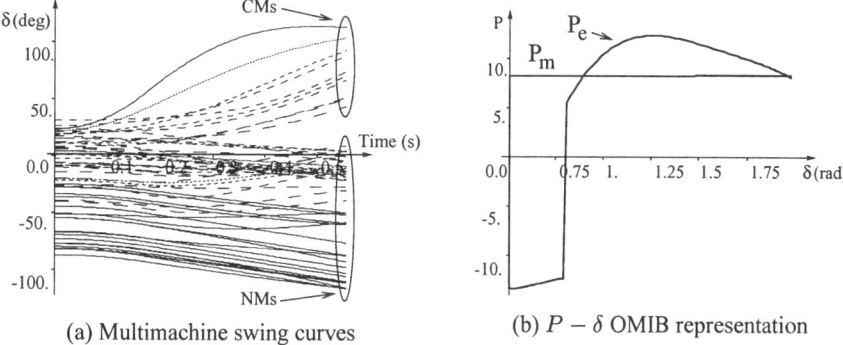

(a) Multimachine swing curves (b) $P - \delta$ OMIB representation

Figure 5.11. Loss of synchronism by first-swing inter-area mode instability. Contingency 239; $t_e = 167$ ms. Base case operating conditions. Adapted from [Bettiol, 1999]

184 (-22 MW), 186 (-55 MW), 187 (-215 MW), 290 (-6 MW), 291 (-2 MW), 292 (-7 MW), 293 (-5 MW), and 2706 (+16 MW).

Thus, recalling that SIME monitors only the contingencies which have the most severely disturbed CMs, i.e., the CMs which require the largest generation shifting, we see that contingency Nr 140 won't be monitored, despite it is initially quite severe.

2.3.5 NMs' generation rescheduling via OPF

Iteration 1. As in the previous case study, the OPF program is run in order to reallocate automatically and "optimally" generation power on NMs, i.e., so as to meet the operating constraints and, at the same time, to maximize the power transfer between exporting and importing areas.

The stability assessment, summarized in Table 5.9, shows that all harmful contingencies are stabilized after the first generation rescheduling.

However, contingency Nr 167 has been significantly over-stabilized (more than 5 % of the maximum generation capacity of the corresponding CM). Hence, the method initiates a second iteration.

Iteration 2. A generation power increase of 134 MW is thus imposed on machine 21. A new OPF is then run for reporting this generation power change on NMs. With the new operating state thus defined, SIME determines a new stable margin for contingency Nr 167, and consequently a small generation power increase to cancel out the stability margin. The convergence criteria being now satisfied, the generation rescheduling iterative procedure is finished.

162   TRANSIENT STABILITY OF POWER SYSTEMS

Table 5.9. Simultaneous stabilization of harmful contingencies. Adapted from [Bettiol, 1999]

1	2	3	4	5	6	7
Base Case						
Contingency Nr	η_2	t_u(MIP) (ms)	CCT (ms)	Swing	ΔP_C (MW)	CM_i (d_i (°), ΔP_{Ci} (MW))
263	-9.30	390	125	+1	-333	187(92,**-215**), 186(82,**-55**), 179(55,**-63**)
140	-2.46	605	142	+1	-185	16(75,-185)
112	-2.99	580	141	+1	-252	16(64,**-252**)
239	-3.49	550	156	+1	-297	181(105,**-115**), 184(95,-22), 182(83,**-46**), 292(79,-7), 187(65,-55), 186(61,-15), 179(57,-24), 291(42,-2), 290(41,**-6**), 293(34,**-5**)
159	-82.31	2,805	157	-3	+16	2706(-13,**+16**)
167	-0.33	2,020	166	+2	-139	21(76,**-139**)
Iteration 1						
263	1.53	(4,000)	–	–	+47	187(–,+30), 186(–,+8), 179(–,+9)
112	1.49	(4,000)	–	–	+84	16(–,+84)
239	3.01	(4,000)	–	–	+248	181(–,+97), 184(–,+18), 182(-,+38), 292(–,+6), 187(–,+46), 186(–,+13), 179(–,+20), 29(–,+1), 290(–,+5), 293(–,+4)
159	37.79	(4,000)	–	–	-5	2706(–,-5)
167	9.76	(4,000)	–	–	+134	21(–, +134)
Iteration 2						
167	5.84	(4,000)	–	–	+4	21(–,+4)

Final Check #1. As already pointed out, a final check should be mandatory for contingencies initially identified by FILTRA as harmful and potentially harmful (in our case, 6+6, see Fig. 5.8).

Thus, the 12 contingencies undergo again the FILTRA procedure, which identifies that contingency Nr 140 is still harmful. Actually, what happened is a change in CM: machine 16 (initially the CM for this contingency) was stabilized after the generation shifting, while machine 2101 becomes now the CM responsible for the loss of synchronism (driven by a backswing multiswing instability).

Table 5.10 summarizes the stabilization procedure of this contingency.

Table 5.10. Iterative stabilization of contingency Nr 140

1	2	3	4	5	6
\multicolumn{6}{c}{Iteration 3 (after final check 1)}					
Contingency Nr	η	sTDI	Swing	ΔP_C (MW)	CM(s) (d(°), ΔP_C(MW))
140	-3.14	3.21	-3	+9	2101 (-137,+9)
\multicolumn{6}{c}{Iteration 4}					
140	5.46	4.00	–	-5	2101 (–,-5)

Table 5.11. Interface flow changes: logical rules vs OPF. Adapted from [Bettiol, 1999]

1	2	3	4	5	6	7	8	9
	\multicolumn{7}{c}{Interface flows (MW)}							
Case	765 kV	\multicolumn{3}{c}{230 kV}			\multicolumn{2}{c}{138 kV}		Total
	110-109	222-2559	202-2603	202-2622	303-2946	337-2608	337-2837	
Initial	2,368	178.0	158.0	129.5	342.0	248.5	121	**3,545**
Final Logical rules	2,056.6	252.5	203.7	143.6	327.6	233.3	119.6	**3,337**
OPF	2,184	257.6	206.8	145.6	326.8	235.8	111.8	**3,468**

Final Check #2. Both, the mandatory and optional checks conclude that all contingencies have been properly stabilized [Bettiol, 1999].

Interface power flows. Table 5.11 compares the OPF results with those obtained by the logical rules reported in [Bettiol, 1999]. As can be seen, the logical rules require 5.87%, while OPF only 2.17% of interface power decrease: obviously, as concerning the MAT problem, OPF is superior to the logical rules [Bettiol, 1999].

Further, OPF is found to be superior also with respect to total losses.

Reliability. The reliability of the results was again checked and found to be fully satisfactory.

Table 5.12. Computational requirements of MAT+OPF

Task Nr	Task name	Computational requirements		
		sTDI	seconds	%
1	**OPF (base case)**	**0.24**	**6**	**0.14**
2	**FILTRA**			
	Filtering block			
	First sub-block (SM)	5.04	126	2.84
	Second sub-block (DM)	23.19	585	13.18
	Assessment block	54.95	1,503	33.87
	Total FILTRA	**83.18**	**2,214**	**49.89**
3	**Stabilization**	**28.72**	**788**	**17.76**
4	**Final check**			
	Harmful and potentially harmful ctgs	38.53	1,021	23.01
	Remaining contingencies	16.22	408	9.20
	Total final check	**54.75**	**1,429**	**32.21**
	TOTAL	**166.89**	**4,437**	**100**

Computational requirements. Table 5.12 summarizes the computing times required by the various tasks. It is worth observing that the overall procedure is little less demanding here than in the first study (4,437 s vs 4,606 s).

This seemingly surprising result may be explained by the fact that the lion's share is taken by FILTRA. And since FILTRA is less demanding here (list L3 contains 18 contingencies vs 30 in the previous study), stabilization influences CPU marginally, even if its size is significantly larger.

2.4 Concluding remarks

The simulations conducted so far suggest the following observations.

- Comparing the TSA&C software interfaced with OPF and with logical ("pragmatic") rules shows that the former is superior. Indeed:
  - the TSA&C + OPF approach yields safe operating conditions closer to the transient stability limits. This permits a larger power transfer limit between exporting and importing areas;
  - use of OPF guarantees that after a generation rescheduling all pre-fault operating constraints (i.e., bus voltages, thermal limits of lines and transformers, etc.) are still met;
  - since the operating constraints of the power system are met by the OPF program, the resulting generation rescheduling pattern has less active power losses.

- The transient stability-constrained maximum allowable transfer (MAT) problem receives a straightforward and systematic solution thanks to the panoply of techniques provided by SIME.
- Calling upon an OPF program suitably adjusted allows providing near optimal solutions to the MAT problem while complementing the target with other objectives:
    - in short, the MAT-OPF-based approach consists of rescheduling power on CMs according to SIME and on NMs according to OPF;
    - computationally, the resulting software can be made fully compatible with real-time requirements because the most time-consuming tasks are straightly parallelizable. The complete iterative cycle "contingency filtering-assessment-stabilization" according to a given objective can thus reach this objective very easily within minutes.
- Further, observe that
    - all harmful contingencies identified by FILTRA are stabilized simultaneously in few iterations;
    - the size of the problem affects rather marginally computational performances.
- Finally, we mention that
    - MAT is an actual problem in many real-world power systems;
    - the proposed solution is likely to be easily accepted by operators, since its strategy and objectives meet everyday concerns;
    - use of an OPF algorithm goes along emerging needs and trends.

3. TSA&C IN CONTROL CENTERS

N.B. This Section is transcribed from [Avila-Rosales et al., 2000].

3.1 Introduction

In recent years utility companies have being required to functionally separate their power transmission, generation and energy marketing departments. Today, utilities are searching for tools in the new restructuring environment that operate at the same efficiency level as the full Energy Management System (EMS) software that has evolved over time. New static and dynamic security challenges are thus emerging.

Independent power producers, competitive economic transactions, open access requirements and free energy routing through the transmission system are putting a greater emphasis on On-line TSA&C, integrated available transfer capability (ATC) static and dynamic calculations and preventive control to allow room for future transactions and savings.

166   TRANSIENT STABILITY OF POWER SYSTEMS

Most companies with stability concerns are willing to consider on-line implementation of TSA&C and the determination of preventive control actions for the operation of their power system.

Below we describe functional aspects for on-line TSA&C integration with the EMS, and the Dispatcher Training Simulator (DTS) environments for planning and current operation studies.

Voltage Security Assessment (VSA) is not of concern here; it is however included in all figures to complete the EMS operation and control center picture.

The following TSA&C/EMS issues are discussed below:

- TSA&C and the Transmission Services Provider (TSP)
- TSA&C and the Independent System Operator (ISO)
- TSA&C for the above considering the study and DTS environments
- ATC calculation considering dynamic security constraints
- preventive control using an optimization technique.

Regarding time horizon, TSA&C must be evaluated to provide ATC and preventive countermeasures for the following:

- planning horizon (hours, days, weeks, months)
- operating horizon (half an hour to one hour).

3.2   TSA&C in the EMS

The TSA&C role in the EMS depends on the type of company. Two basic alternatives are described here.

3.2.1   On-line TSA&C for the TSP

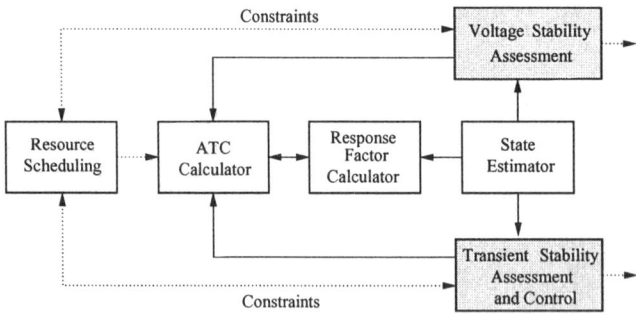

Figure 5.12.   Transmission Services Provider. Adapted from [Avila-Rosales et al., 2000]

Figure 5.12 is a functional overview of the TSP environment. The purpose of the ATC calculator is to manage available transfer capability information for the OASIS (Open Access Same time Information System) automation system.

It calculates initial ATC values during periods of initialization or resynchronization. It also calculates changes to ATC values that result from new requests for transmission service. The ATC calculator in this context currently performs either a flow based or a path based method.

Figure 5.12 shows that the ATC calculator supports data input from multiple sources: the response factor calculator using state estimation information, the resource scheduling, TSA&C and VSA if available.

The ATC calculator evaluates ATC data for each constrained facility based on the operating horizon (solid lines), the planning horizon (dotted lines) and in some cases a study horizon.

When a request for transmission capacity is received, the ATC calculator calculates new ATC data and sends the values for evaluation. Once evaluated, the new ATC values may be posted. ATC values are calculated in advance and for future transmission reservations considering two horizons:

- *the operating horizon* which starts in the next hour and covers the immediate short term;

- *the planning horizon* which starts after the operating horizon and extends into the future.

Transfer paths are used to evaluate stability limits and the information is sent to the ATC. A fast technique is required to update the limits when a specific request for transmission capacity reservation is received.

Response Factor (RF) runs in real-time using fresh information from state estimation; it currently calculates flow at each gate and the sensitivity of each flow gate to a new reservation on a given path. TSA sensitivity values will be considered in the RF evaluation for the constrained facilities (related to the transfer path) to obtain ATC estimated values for future transactions.

Resource scheduling uses resource commitment network data, forecasted conditions, transaction information and device schedules to optimize and obtain transfer limits. TSA&C will include stability constraints as part of the calculation.

VSA and TSA may reside in independent boxes and use distributed processing to minimize execution time.

3.2.2 On-line TSA&C for the ISO

Figure 5.13 describes an ISO environment using a resource dispatch that provides solutions for the market based operation.

TSA&C will include stability constraints in the market solvers for the required time horizon and interfaces with the resource commitment, the resource scheduling and the resource dispatch applications.

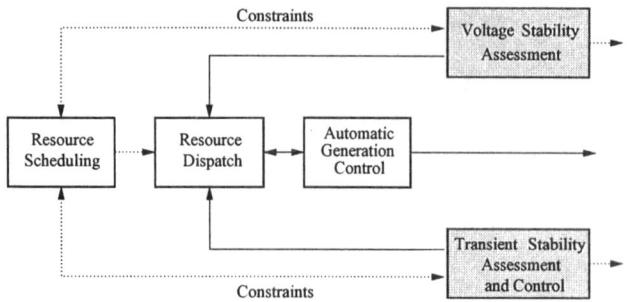

Figure 5.13. Independent System Operator. Adapted from [Avila-Rosales et al., 2000]

Resource scheduling is a function with a time horizon of 24 hours in hourly intervals. The results are fed into the EMS market database and TSA&C and resource dispatch will iterate to ensure dynamic security in this time frame.

Resource dispatch executes typically every five minutes and its purpose is the real-time dispatch of the resources; it uses generator response and limits to calculate base points for AGC as well as unit participation factors.

Note the planning horizon (dashed line) and the current operating loop (solid line) are connected with TSA&C to include stability constraints.

TSA&C can provide limits every five minutes, with a fast and reliable sensitivity approach.

Several harmful dynamic contingencies and transactions are evaluated for real-time and planning scenarios.

For the planning horizon, TSA&C will analyze contingencies and transactions for daily operation and up to two weeks in advance.

TSA&C and VSA may reside in independent boxes and use distributed processing to minimize execution time.

3.3 Congestion management

In general the EMS should be provided with a set of optimization tools that will allow Congestion Management (CM) and generation redispatch. Preventive control is an example where TSA&C and these optimization tools can be used. The integration of the planning and the current operating horizons will force the extensive use for next hour, day and month. TSA&C shall interface with these tools to analyze dynamic contingencies, then stabilize unstable cases, possibly via OPF.

The CM process will participate in activities like ATC calculation and response against security violations. These tools may be used to plan and schedule controls for dangerous situations. The controls will have of course associated cost of operation; but in some cases they could be cost free.

CM tools shall be flexible and easy to use by different applications like TSA&C and VSA to consider transient and voltage problems. Resource scheduling and resource dispatch are part of the applications including optimization tools in new restructuring EMS.

3.4 TSA&C for the DTS and Study Environments

The DTS provides a realistic environment for operators to practice normal, everyday operating tasks and procedures, as well as emergency conditions. It can be used in an experimental and investigatory manner to recreate past scenarios, or simulate future behavior of the system and the EMS. The DTS and Study environments should include resource scheduling, resource dispatch, VSA and TSA&C to emulate the new energy system. Remedial actions can be integrated as part of the simulation process, thereby providing a powerful dynamic tool. The following features are required:

- The DTS fast real-time initialization
- The Data Preparation (DP) using the DTS databases
- Network information, transactions, contingencies, monitored elements and parameters are sent to the TSA&C box
- TSA&C output is summarized and sent to the DTS to continue with the near future base case condition.

The DTS can be thought of as being parallel to, and a close emulation of the EMS environment. Hence, results in the DTS environment are highly reliable and accepted by the EMS users.

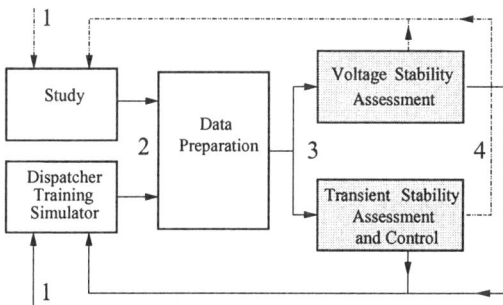

Figure 5.14. Study and DTS dynamic tools. Adapted from [Avila-Rosales et al., 2000]

The study tools are shown with dashed lines in Fig. 5.14; the loop is similar to the DTS one, except that is primarily used for power flow case analysis including optimal power flow and contingency analysis. Study is initialized from real-time or saved cases to recreate actual conditions and analyze the impact of "what if" conditions to prevent dangerous situations.

4. SUMMARY

This chapter has set up a unified TSA&C software, based on material developed in the previous chapters.

Section 1 has first described the resulting basic software then has interfaced it with an OPF algorithm.

Section 2 has used this augmented "TSA&C + OPF" software to assess transient stability-constrained maximum power transfer between areas of the Brazilian South-Southeast-Centerwest power system. This assessment relies on generation shifting, where the SIME-based TSA&C software takes care of generation rescheduling of critical machines, whereas the OPF software takes care of generation rescheduling of non-critical machines. It was shown that this integrated software provides "near optimal" solutions able to comply with real-time operation requirements.

Besides, the OPF algorithm broadens the possibilities of TSA&C by determining static limits as well. Even more importantly, it bridges the gap with other EMS functions. Its role as the ISO's tool for safeguard and justification in his interaction with the market is certainly not the least interest of its interface with TSA&C.

ISO's concerns have been addressed in Section 3, and various solutions were envisaged for the combined coverage of dynamic (transient and voltage stability) constraints together with static ones, in particular during congestion management and ATC calculations.

Chapter 6

CLOSED-LOOP EMERGENCY CONTROL

The previous chapters have addressed preventive TSA&C issues and developed appropriate techniques. In particular, Chapters 4 and 5 focused on preventive control aiming to design pre-contingency remedial actions able to stabilize harmful contingencies, should they occur. The actions concerned generation shifting.

This chapter addresses emergency control issues. It aims at designing, in real time, corrective post-contingency actions, triggered during the transient period following a contingency inception, so as to avoid loss of synchronism which otherwise would occur. The corrective actions concern generation tripping, although many other types of control may also be thought of.

More precisely, the objectives of this chapter are:

- *to predict in real-time, after a disturbance inception, whether the system is driven to instability*

- *if yes, to assess the size of the instability and devise "in extremis" corrective actions able to safeguard system's integrity*

- *to continue monitoring the system in order to assess whether the action has been sufficient or should be further complemented.*

In short, the ultimate objective is to design techniques for real-time closed-loop transient stability emergency control. The information necessary to achieve such a challenging task should be provided by real-time measurements rather than T-D simulations.

1. OUTLINE OF THE METHOD
1.1 Definitions

Transient stability emergency control may be viewed as complementary to preventive control in many respects. For example, preventive control deals with pre-contingency actions, emergency control with post-contingency ones. Also, preventive control attempts to broaden the system domain of attraction so as to contain the dynamic trajectory initiated by a contingency inception. On the other hand, emergency control attempts to bend the dynamic trajectory so as to force it to remain or return to the domain of attraction (which will generally be changed, too). Differences between preventive and emergency control are further discussed in Chapter 7.

The preventive countermeasure considered in Chapters 4 and 5 for real-time **preventive control** was generation shifting, possibly in conjunction with load shedding[1]. The decision about whether to take such countermeasure(s) or not relies on the tradeoff between economy and security.

Emergency control, on the other hand, aims at designing and triggering actions in real time, *after* a harmful contingency *has actually occurred*. Here, such control actions become vital for both security and economy.

Emergency actions may be either designed in real time using real-time measurements, or assessed in anticipation, for example by means of off-line stability simulations. The latter case belongs to **open-loop emergency control**, as opposed to **closed-loop emergency control**; in this latter case, the action *is designed and triggered in real time, during the transient period following a contingency inception, and the system continues being monitored and further controlled.*

1.2 Scope

The Emergency SIME proposed in this chapter is a general approach to real-time closed-loop transient stability emergency control [Zhang et al., 1997b, Ernst et al., 1998b, Ernst and Pavella, 2000, Ernst et al., 2000a].

Unlike Preventive SIME which goes along the conventional thinking and strategies[2], Emergency SIME departs definitely from the traditional T-D simulation-based approaches by processing real-time measurements.

Admittedly, given that transient stability phenomena develop very fast, the above objective seems to be quite ambitious with respect to both software and hardware requirements. This may explain to a large extent why, to the best of our knowledge, such a challenging strategy has not been envisaged so far.[3]

---

[1] although in the simulated real-world examples load shedding has not been required.
[2] even if its possibilities are far beyond those of conventional transient stability methods
[3] Note, however, that open-loop emergency control schemes exist, in some dedicated places.

We however trust that the proposed method is now within reach, thanks to the software possibilities of SIME and to recent important technological advances. In what follows we address methodological issues essentially.

Let us observe at once that, whatever the type of control, the closed-loop emergency control techniques developed hereafter are intended to safeguard important, dedicated sites as, for example, large hydro-electric power plants. They are supposed to be worked out at dedicated locations, different from the control room.

1.3 Principle

As aforementioned, the Emergency SIME uses measurements acquired from the system power plants in real-time, in order to control the system *just after* a contingency occurrence and its clearance.

More precisely, following a disturbance inception and its clearance, Emergency SIME aims at predicting the system transient stability behavior and, if necessary, at deciding and triggering control actions early enough to prevent loss of synchronism. Further, it aims at continuing monitoring the system, in order to assess whether the control action has been sufficient or should be reinforced. More precisely, the procedure consists of the following tasks.

(i) *Predicting the OMIB structure*: say, 100 ms ahead.
(ii) *Predicting the OMIB $P - \delta$ curve*, using a weighted least-squares (WLS) estimation.
(iii) *Predicting instability*, by searching whether the above curve reaches SIME's instability conditions.

If not, repeat the above steps using new measurement sets. If yes, compute the corresponding margin, as well as the (predicted) time to instability.

(iv) *Determine the size of control* and trigger the corresponding action.
(v) *Continue monitoring* by repeating above steps after the actions have properly been triggered.

N.B. *Prediction's validity test*. The reliability of the prediction takes advantage of the observation that, since the operating and contingency conditions are fixed, the value of the (negative) margin should be constant, whatever the time step (see below, § 2.2). Hence, the above computations should be repeated at successive Δt's until getting an (almost) constant margin value.

1.4 General organization

Figure 6.1 describes schematically the general framework for real-time transient-stability closed-loop emergency control.

Let us briefly comment on the functions contained in the various boxes.

174   TRANSIENT STABILITY OF POWER SYSTEMS

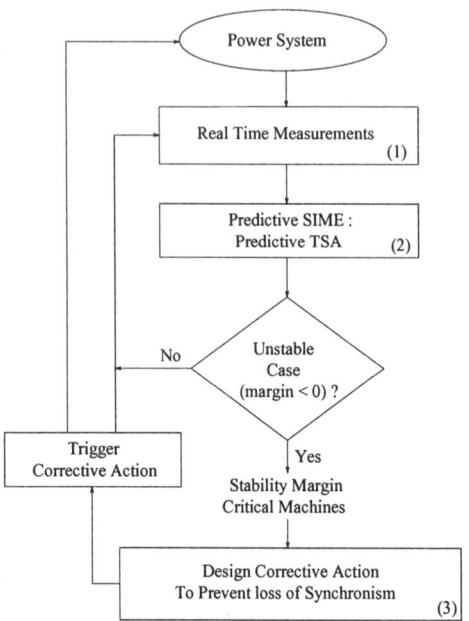

Figure 6.1. General organization of real-time transient stability closed-loop emergency control. Adapted from [Ernst and Pavella, 2000]

(1) **Real-time measurements**. These measurements are supposed to be collected at main power plants and centralized in a location (control room) possibly close to the dedicated site under control. The measured (directly or indirectly) quantities are machines rotor angles, speeds and accelerations.

(2) **Predict TSA**: see Section 2.

(3) **Design and trigger corrective actions**: see Section 3.

1.5   Computational issues

The sine qua non of method's practical interest is its ability to prevent loss of synchronism in real-time. Given the rapidity of transient (in)stability phenomena evolution, the duration of a complete emergency control cycle should not exceed a few hundreds of milliseconds.

Let us appraise this duration, and more specifically, the time elapsed between a contingency clearance and the moment the corrective action starts acting. To this end, we will first consider the various tasks of the emergency control (EC) scheme displayed in Fig. 6.1, then we will appraise their corresponding durations, taking into account performances that one can reasonably expect from modern telecommunication and measurement devices.

1.5.1 Involved tasks

According to Fig. 6.1, an EC cycle comprises the following tasks:

(i) data acquisition at power plants and their transmission to the control room
(ii) data processing at the control room (blocks (2) and (3) of the figure)
(iii) order transmission from the control room to the power plant(s) to be controlled
(iv) order actual application.

1.5.2 Corresponding durations

1. For the data acquisition-transmission corresponding to above item (i): 50 ms
2. For above item (iii): 50 ms
3. For above item (iv): 50 ms
4. To appraise the duration of task (ii), note that the data processing for predicting (in)stability and computing the corresponding margin requires a minimum of 3 successive measurement sets and up to, say, 10 sets. Assuming that the rate of data acquisition-transmission is about 1 cycle (to simplify, say, 20 ms) this corresponds to a total duration of 60 to 200 ms.
5. The time to run the corresponding software is virtually negligible with respect to the above durations.

Summing up the above approximate figures yields a total duration varying in between 210 ms and 350 ms after the contingency clearance, i.e., in between 310 and 450 ms after the contingency inception. For the EC scheme to be effective, these durations should be smaller than the time to instability, t_u, of the corresponding contingency.

1.6 Notation specific to Emergency SIME

In addition to the general acronyms and notation used so far, in this chapter we will further introduce the following specific ones.

$t_0 = 0$: beginning of the during-fault period
t_e : beginning of the post-fault period
Δt : sample time, i.e. time between two successive measurement sets acquisition (here: $\Delta t = 20$ ms)
t_f : beginning of the predictive TSA
t_i : current processing time
t_{ct} : control time, i.e., time elapsed between a contingency inception and the control action. (Subscript ct is used in order to avoid any confusion with t_c, the critical clearing time.)
t_d : sum of durations of steps (i), (iii) and (iv) of § 1.5.1
$\delta_i = \delta(t_i)$: OMIB angle at t_i
$\omega_i = \omega(t_i)$: OMIB speed at t_i.

176 TRANSIENT STABILITY OF POWER SYSTEMS

2. PREDICTIVE SIME

2.1 Description

Unlike preventive TSA, the predictive TSA has not been used so far, because this task is not achievable by conventional approaches; besides, its interest is linked to the feasibility of closed-loop emergency control - and, again, this cannot be handled by conventional approaches.

Predictive TSA deals, in real-time, with an event (or a succession of events) which has been *detected but not necessarily identified*, and generally automatically cleared by the protective devices. Thus, in order to be effective, it must *predict* the system behaviour *early enough* so as to leave sufficient time for determining and triggering appropriate control actions, whenever necessary. To get a stability diagnostic ahead of time, the predictive TSA relies on real-time measurements.

More precisely, the method predicts the stability of the system entering its post-fault configuration, using the multimachine data available at successive sample times Δt's (e.g., 1 sample every 20 ms). Thus, at each sample time, an OMIB analysis is performed to decide whether the system keeps stable or is driven to instability. The crux for this analysis is the prediction of the OMIB $P_a - \delta$ curve, and hence the prediction of the unstable angle, δ_u, and the corresponding stability margin. Its achievement addresses the following two questions:

1. which are the most disturbed machines ?

2. is the system driven to (in)stability and to what extent ?

2.2 Procedure

To answer the above questions the method relies on the following steps, illustrated in Figs 6.2.[4]

(i) At a time t_i short after the disturbance clearance, $(t_i \geq t_e + 2\Delta t)$ consider the incoming measurements at times $t_i - 2\Delta t$, $t_i - \Delta t$, t_i, and use Taylor series to predict the individual machine angles at some time ahead (e.g. 100 ms). Sort the machines in decreasing order of these angles and consider as candidate critical machines those advanced machines which are above the largest (angular) distance between two successive machines.

(ii) Construct the corresponding OMIB, determine its parameters $(\delta, \omega, \gamma, P_a)$ from the corresponding parameters of the individual power plants at times

---

[4]Figure 6.2a sketches the principle, while Fig. 6.2b illustrates its application to the real case simulated in Section 4. Notice that the curves in Fig. 6.2b are drawn after the disturbance clearance. (Actually, they start being drawn 10 time samples after the first acquisition of measurement set.)

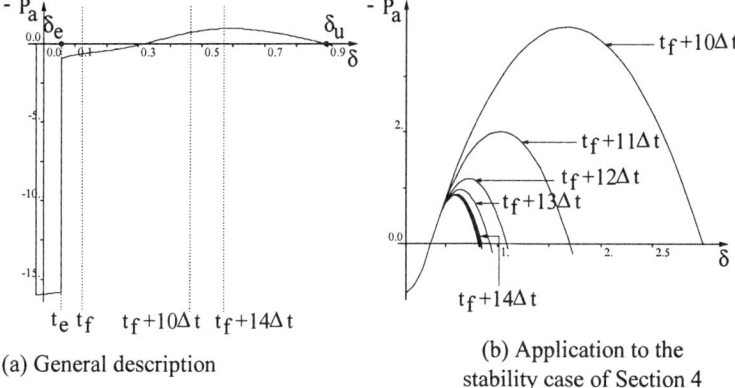

Figure 6.2. Principle of the predictive SIME

$t_i - 2\Delta t$, $t_i - \Delta t$, t_i, and approximate the $P_a - \delta$ curve by solving:

$$\hat{P}_a(\delta) = a\delta^2 + b\delta + c \qquad (6.1)$$

for a, b, c at these times.[5].

(iii) Solve eq. (6.1) to find the OMIB angle $\delta_u > \delta(t_i)$ which verifies conditions (2.15).

(iv) Compute the stability margin, η, according to (2.14), (2.18), (2.19):

$$\eta = -\int_{\delta_i}^{\delta_u} P_a d\delta - \frac{1}{2} M\omega_i^2 . \qquad (6.2)$$

(v) If η is found to be negative or close to zero, declare the system to be unstable and determine control actions (see Section 3).

(vi) Compute the *time to instability*, t_u, i.e. the time for the OMIB to reach its unstable angle, δ_u, i.e. to go unstable. This may be computed, for example, by [Ernst and Pavella, 2000]:

$$t_u = t_i + \int_{\delta_i}^{\delta_u} \frac{d\delta}{\sqrt{\frac{2}{M} \int_{\delta_i}^{\delta} -P_a d\delta + \omega_i^2}} . \qquad (6.3)$$

(vii) Acquire a new set of measurements and continue monitoring the system.

---

[5] Subsequently, this estimated value, $\hat{P}_a$, is refined by using newly acquired sets of measurements and processing a least squares technique which shows to be particularly robust. A further improvement consists of using a weighted least-squares (WLS) technique, by giving more important weights to the last sets of measurements. To simplify notation, however, we will simply write P_a, even if all values are actually estimated.

2.3 Remark

Two main ideas are behind the above predictive stability assessment: OMIB structure and prediction's validity test.

- OMIB structure. The OMIB used at the very first instants of the procedure relies on measurement-based prediction rather accurate assessment. Therefore, it might not necessarily be fully correct. However, it is likely to contain (part of) the most disturbed machines, whose control will (hopefully) stabilize the system.
- Validity test. The resulting $P_a - \delta$ curve might be not accurate enough. But its accuracy may be assessed by observing that, by definition, for a fixed clearing time t_e the margin (6.2) should be constant whatever t_i ; hence, the margin values obtained at successive t_i's should converge to a (nearly) constant value. This observation provides an interesting, handy validity test.

Figure 6.2b shows that the $P_a - \delta$ prediction converges towards the exact $P_a - \delta$ following about 14 time samples after the first prediction; this corroborates what Table 6.1 of Section 4 shows: the value of η stabilizes at about 435 ms.

2.4 Specifics

1.- Computationally, the above strategy is very inexpensive and fast; indeed, at each time sample, it merely requires:

(a) solving Taylor series for the individual machines to identify the candidate OMIB;
(b) computing this OMIB parameters and its P_a curve (6.1);
(c) solve (6.1) to get δ_u ;
(d) compute the margin (6.2).

Obviously, all these computations require only fractions of ms.

2.- The time to instability, t_u , expressed by (6.3) is a good indicator of contingency severity; moreover, it provides valuable advice about whether to act immediately, though imperfectly, or to wait for a more accurate assessment.

3.- It may happen that the transient stability phenomena take some time to get organized, and do not appear clearly enough at the beginning of the post-fault transients, thus yielding a confused diagnostic. However, in such cases instability is likely to develop rather slowly; this leaves time to continue monitoring until the phenomena become clearer. (See Table 6.1 of Section 4.)

4.- Along the same lines, a case which at the first time prediction yields a stable margin may actually be unstable. The closed-loop control handles properly such cases.

5.- The above developments assume that the individual power plant variables may be obtained by synchronized phasor measurement devices placed at each power plant together with some local processing power to determine generator angles, speeds and accelerations.

2.5 Salient features

- The prediction phase starts after detecting an anomaly (contingency occurrence), generally followed by its clearance via protective relays. Note that this prediction does not imply identification of the contingency (location, type, etc.).

- The prediction is possible thanks to the use of the OMIB transformation; predicting the behavior (accelerating power) of all of the system machines would have led to totally unreliable results.

- There may be a tradeoff between the above mentioned validation test and the time to instability: the shorter this time, the faster the corrective action should be taken. On the other hand, the severer the contingency, the earlier the instability phenomena appear.

3. EMERGENCY CONTROL
3.1 General principle

As already stated, stabilizing an unstable case consists of canceling out the negative margin, i.e. of increasing the decelerating area and/or decreasing the accelerating area in the OMIB $P - \delta$ plane (see Fig. 6.2a).

Broadly, this may be achieved:

- either by reducing the mechanical power of the OMIB[6] and hence of the CMs. E.g., by using:
  - fast-valving, or
  - generator tripping;

- or by increasing the electrical power. E.g., by using:
  - dynamic braking
  - HVDC links
  - thyristor controlled series, and other FACTS devices.

Further, notice that a negative margin means that the integral term in eq. (6.2) is not large enough: in order to stabilize the system one should increase this area by increasing the decelerating power.

---

[6]unless backswing phenomena are of concern, in which case the CMs' power should be increased.

Finally, recall that, in addition to the time needed to predict the unstable margin, there is always a time delay, t_d, before the corrective action is triggered; it corresponds to the three terms (i), (iii) and (iv) of § 1.5.1. Observe that the longer the time delay, the larger the size of the corrective action needed. These issues are addressed below.

3.2 Generation shedding

3.2.1 Computing stability margins

The aim of this paragraph is to assess the influence of generation shedding on the system stability margin. We will start by considering the shedding of one critical machine, denoted m_j. It is supposed to be shed x seconds after the current time t_i.[7] Note that the resulting procedure can easily be extended to any number of machines, shed at different times.

We assume that at time t_i, n sets of measurements corresponding to the post-fault period have already been acquired, where n is at least equal to 3. These sets correspond to times t_i, $t_i - \Delta t$, ..., $t_i - (n-1)\Delta t$. The critical OMIB identification described in previous sections relies on the values of parameters $(\delta, \omega, \gamma, P_a)$ computed from the n sets of measurements according to the procedure described in § 2.2.

Shedding machine m_j, x seconds after the last set of measurement acquisition, results in modifying OMIB's structure, since the number of CMs decreases by one.

Hence, the first task is to predict the angle and the speed of this OMIB just after the actual shedding of machine m_j. Let OMIB$^{(1)}$ denote this new OMIB. To this end we first compute OMIB$^{(1)}$ from the n sets of measurements, using eqs (2.3) to (2.12) of Chapter 2 where C is replaced by $C \setminus \{j\}$ to indicate that machine m_j does not anymore belong to the group of CMs. Superscript $^{(1)}$ distinguishes the parameters of this new OMIB$^{(1)}$, from the original parameters.

Accordingly, we use the n sets of parameters to approximate the $P_a^{(1)} - \delta^{(1)}$ curve by solving
$$P_a^{(1)} = a^{(1)}\delta^2 + b^{(1)}\delta + c^{(1)}.$$

The angle that OMIB$^{(1)}$ reaches at the control time, i.e., x seconds after the current time, is denoted $\delta_{ct}^{(1)}$ and computed using the following equation:

$$x = \int_{\delta_i^{(1)}}^{\delta_{ct}^{(1)}} \frac{d\delta^{(1)}}{\sqrt{\frac{2}{M^{(1)}} \int_{\delta_i^{(1)}}^{\delta^{(1)}} -P_a^{(1)} d\delta^{(1)} + (\omega_i^{(1)})^2}}. \quad (6.4)$$

---

[7]We choose x instead of t_d to emphasize that whatever the time horizon of the shedding, we can estimate the influence of this control action on the system stability.

Chapter 6 - CLOSED-LOOP EMERGENCY CONTROL

Note that solving this equation for $\delta_{ct}^{(1)}$ can be done only numerically.

Once $\delta_{ct}^{(1)}$ is computed, the value of the OMIB$^{(1)}$ speed at the control time can be determined by solving the following equation for $\omega_{ct}^{(1)}$:

$$-\frac{1}{2}M^{(1)}\omega_{ct}^{(1)2} = -\int_{\delta_i^{(1)}}^{\delta_{ct}^{(1)}} P_a^{(1)} d\delta^{(1)} - \frac{1}{2}M^{(1)}\omega_i^{(1)2}. \quad (6.5)$$

Note that the only approximation used to compute $\delta_{ct}^{(1)}$ and $\omega_{ct}^{(1)}$ is the extrapolation of the $P_a^{(1)}$ curve.

To compute the stability margin of the corrected system we still need the shape of the accelerating power of OMIB$^{(1)}$. Note that the $P_a^{(1)} - \delta^{(1)}$ curve previously computed is not appropriate for this purpose. Indeed, equality $P_a^{(1)} = P_m^{(1)} - P_e^{(1)}$ is valid under the assumption that machine m_j is still in activity. But, obviously, while the shedding of machine m_j does not influence $P_m^{(1)}$ (at least at the very first moments after this shedding), it does influence $P_e^{(1)}$.

To simplify, we will approximate this influence by considering that the electrical power produced by the critical machines is a function of the angle of the OMIB of concern, independent of the number of machines still in activity. The same assumption will be used for the non-critical machines. [8] So we will suppose that the electrical output produced by the machines of group $C\setminus\{j\}$ just after t_{ct} equals the electrical output produced by the machines of group C just before t_{ct}. We may express this in a different way, by writing: $\sum_{i\in C} P_{e_j}$ just before t_{ct} equals $C\setminus\{j\}$ just after t_{ct}.

Now, to compute the shape of the accelerating power of OMIB$^{(1)}$ after the corrective time[9], formulas (2.3) to (2.12) suggest to compute n sets of parameters $(\delta^{(2)}, P_a^{(2)})$ where

$$\delta^{(2)} = \delta^{(1)} \text{ and } P_a^{(2)} = P_a^{(1)} - M^{(1)}\frac{P_{e_j}}{M_C^{(1)}}.$$

Further, we use the n sets of parameters to compute $a^{(2)}$, $b^{(2)}$ and $c^{(2)}$ by solving the equation

$$P_a^{(2)} = a^{(2)}\delta^2 + b^{(2)}\delta + c^{(2)}.$$

---

[8] Note that more sophisticated and reliable models can also be implemented; but they require additional computations.
[9] The parameter of the modified OMIB after the time to control will be identified by superscript $^{(2)}$.

182   TRANSIENT STABILITY OF POWER SYSTEMS

Finally, using the values of $a^{(2)}$, $b^{(2)}$ and $c^{(2)}$ thus computed, we solve the above equation to get the value of the unstable angle of the controlled system, denoted $\delta_u^{(2)}$.

The stability margin of the controlled system is readily given by:

$$\eta = -\int_{\delta_{ct}^{(1)}}^{\delta_u^{(2)}} P_a^{(2)} d\delta^{(2)} - \frac{1}{2} M^{(1)} \omega_{ct}^{(1)2}. \tag{6.6}$$

In Fig. 6.3a, the P_a curve corresponds to the original OMIB accelerating power. In Fig. 6.3b the dotted curve corresponds to the P_a curves of OMIB$^{(1)}$, while the solid curve corresponds to the same OMIB$^{(1)}$ but after the corrective action has been applied.

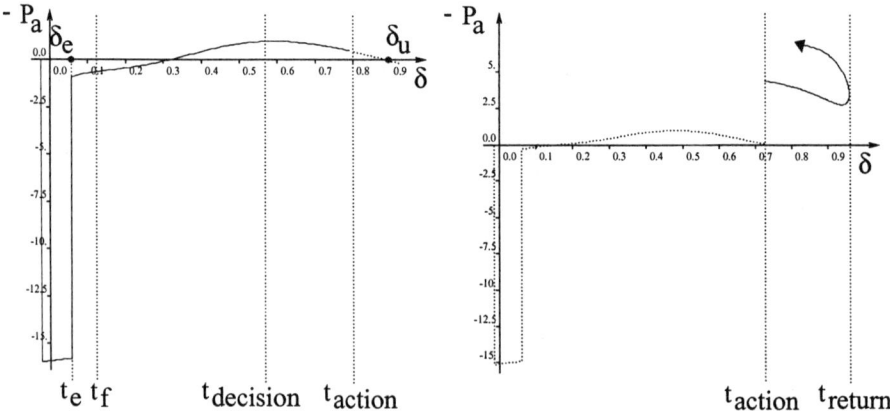

(a) Uncontrolled OMIB $P - \delta$ curve   (b) Controlled OMIB $P - \delta$ curve

Figure 6.3.   EPRI 88-machine test system

3.2.2   Identification of the machine(s) to shed

The next step consists of determining which machine(s) to shed when the system is found to be unstable. The proposed method is simple: from the predicted swing curves of the individual machines, take the most advanced one, and compute the margin of the corrected system using the procedure described in the previous paragraph where x equals t_d; if the margin is still negative, shed another machine; continue until the margin becomes positive.

Chapter 6 - CLOSED-LOOP EMERGENCY CONTROL   183

Once the shedding order is sent, continue monitoring the system. To compute the margin of the controlled system using the next set of measurements, proceed like in to the previous paragraph, after replacing x by $t_d - \Delta t$.

3.2.3 Straightforward improvements

A good number of improvements can be made to what has been described in the two previous paragraphs.

For example, other functions than second order functions can be used to approximate the P_a curves. A function of higher order or of a different type (e.g. $a * cos(b * \delta + c)$) could offer better results.

The method described to choose when and which machines to shed is also perfectible. Indeed, margins close to zero could lead to erroneous corrective actions because of the approximations done.

But all these considerations are beyond the scope of this chapter which merely aims to give a flavor of method's basics.

4. SIMULATIONS
4.1 Description

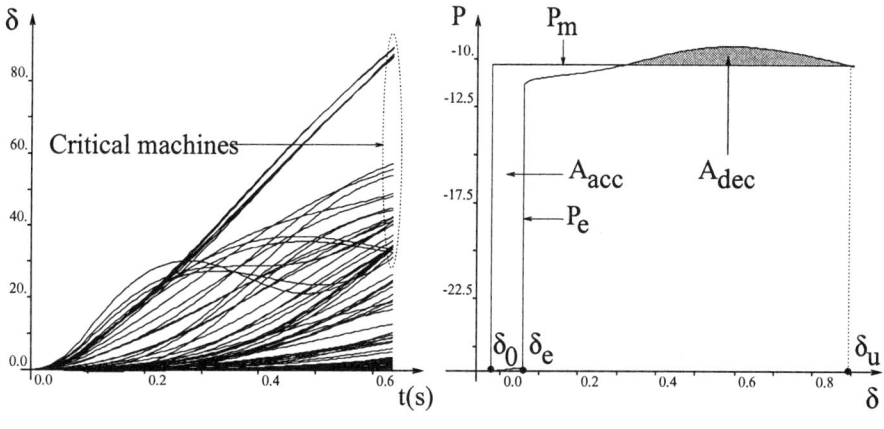

(a) Swing curves   (b) OMIB $P - \delta$ curve

Figure 6.4.   Uncontrolled EPRI 88-machine test system

The simulations are performed on the EPRI test system C, having 88 machines (of which 14 are modeled in detail, see Section 3 of Appendix B). The considered base case has a total generation of 350,749 MW.

184   TRANSIENT STABILITY OF POWER SYSTEMS

The considered contingency is a 3-ϕ short-circuit applied at bus #15 (500 kV); it is cleared 100 ms after its inception, ($t_e = 100$) by opening the line 1-15.

The ETMSP program is used here to create artificially real-time measurements, since such measurements are not available. The swing curves of the uncontrolled system corresponding to this contingency are displayed in Fig. 6.4a. The corresponding $P_a - \delta$ curve is portrayed in Fig. 6.4b.

4.2   Simulation results of Predictive SIME

Table 6.1.   Closed-loop emergency control

1	2	3	4	5
t_i (ms)	δ_u (rad.)	t_u (ms)	η/M (rad/sec)^2	η/M after shedding
375	1.094	788	-0.60	
395	0.922	676	-0.81	
Corrective decision is taken (3 units shed)				
415	0.850	631	-0.88	0.271
435	0.822	614	-0.91	0.115
455	0.813	610	-0.91	0.092
475	0.820	617	-0.91	0.113
495	0.826	622	-0.90	0.151
515	0.836	631	-0.90	0.234
535	0.850	642	-0.89	0.347
555	0.858	649	-0.89	0.376
Corrective action is applied				
575	0.861	652	-0.89	0.352
595	0.860	652	-0.89	0.361
615	0.859	651	-0.89	0.373
635	0.861	652	-0.89	0.384

The simulations of the predictive TSA are displayed in Fig. 6.2b. On the other hand, Table 6.1 summarizes the results of both predictive TSA and closed-loop emergency control.

Simulation conditions :
$t_0 = 0$ (contingency inception);
$t_e = 100$ ms ;
first set of data: acquired at $t_i = 115$ ms; (arbitrarily taken larger than 100 ms);
rate of data acquisition: $\Delta t = 20$ ms .

Simulation results of the predictive TSA. The predictive TSA computations start at $t_f = 115 + 2 \times 20 = 155$ ms (remember, three measurement sets are necessary for running the predictive TSA (§ 2.3)).

At the beginning (155 ms up to 375 ms), the simulations do not provide a clear prediction (identification of CMs and corresponding margin): it seems as though the system is not going to lose synchronism (see also Fig. 6.2b). But at $t_i = 375$ ms (i.e., $t_f + 11 \Delta t$) the first unstable margin appears; it corresponds to 33 CMs.[10]

Table 6.1 summarizes the sequence of events from 375 ms onwards. Observe that the time to instability predicted at $t_i = 395$ ms is quite short; it equals around 670 ms, i.e., less than 300 ms later. This is why a corrective action is decided before waiting for the margin to converge to a constant value, which is (nearly) reached at $t_i = 435$ ms.

Figure 6.2b illustrates the above descriptions; it also suggests that:

- there is no negative margin before $t_i = 375$ ms $(= t_f + 11 \Delta t)$
- the prediction starts being reliable around 435 ms $(= t_f + 14 \Delta t)$; indeed, the $P_a - \delta$ curve drawn at 435 ms is found to coincide with the "exact" curve, obtained by the preventive SIME.

It is interesting to compare results of the preventive and predictive SIME:

- the time to instability, t_u, is found to be 635 ms by the preventive SIME, whereas the predictive SIME yields values varying between 788 ms (at $t_i = 375$ ms) and 614 ms (at $t_i = 435$ ms);
- the normalized margin is found to be -1.044 $(rad/s)^2$ by the preventive SIME, whereas the predictive TSA underestimates it slightly (around -0.9).

4.3 Simulation results of Emergency Control

Because of the proximity to instability, it is decided to take control action quite early (at $t_i = 415$ ms). The type of action is shedding CMs; the size of this action, assessed according to § 3.2, is found to be 3 units among the 7 more advanced ones, corresponding to 2,463 MW.[11]

Table 6.1 summarizes the sequence of the events.

- Let us first focus on columns 2 to 4 which refer to the monitoring of the system, relying on the incoming measurements at the rate of 20 ms: rows $t_i = 415$ to 555 ms correspond to the measurements of the uncontrolled

---

[10] Note that for $t_i = 355$ ms (i.e., $t_f + 10 \Delta t$) the margin is positive and thus the unstable angle represented in figure Fig. 6.2b is never reached

[11] The machines shed are 1875 (835 MW), 1771 (793 MW) and 1877 (835 MW).

system; they suggest that the predicted loss of synchronism reaches good accuracy (the margin value stabilizes around -0.90 (rad/s)^2) and that it is imminent: the t_u value stabilizes around 640 ms. At $t_i = 575$ ms, i.e. 160 ms after the control decision has been taken, the control (generators' shedding) is actually triggered.

- Consider now column 5 of the table which refers to the controlled system, i.e. to the system evolution after the shedding of 3 units. They are all predicted results: rows corresponding to $t_i = 415$ up to 555 ms predict the system evolution as will be after the control triggering, whereas rows corresponding to $t_i = 575$ up to 635 ms assess the system evolution after the control triggering.

Observe that the negative margin of column 4 stabilizes to a more constant value than the positive margin of column 5; this is due to the fact that the negative margin relies on the closed-form analytical expression (6.2), whereas the positive margin results from an approximate expression.

Figure 6.3 describes the sequence of the closed-loop control events: the control decision is taken at $t_i = 415$ ms (corresponding to an angle of about 0.57 rad), relying on stability conditions corrresponding to 33 CMs. The control starts acting at $t_i = 555$ ms. It is found to be sufficient to stabilize the system: the $P_a - \delta$ curve experiences a return angle of about 0.93 rad, corresponding to $t_r = 975$ ms.

5. DISCUSSION

5.1 Summary of method's features

The Emergency SIME uses generator data of the main power plants of a utility, supposed to be derived from synchronized phasor measurement devices placed at each plant, then sent to a central location. The method predicts the transient stability behaviour of the power system, and determines whether and which control actions should be taken in order to stabilize it. These actions are then sent back to the control devices. The total time required for the whole cycle "prediction-assessment-design and triggering of the action" is expected not to exceed 500-600 ms; and the severer the instability, the shorter the duration of this cycle.

An essential asset of this real-time approach is its adaptability to any kind of operating and fault conditions. Moreover, being free from any off-line (or preventive mode) tuning, it is intrinsically robust with respect to modeling errors, and able to cope with unforeseen events.

Another method's asset is its high computational efficiency. This makes it able to work in a closed-loop fashion, using a fast enough ground communication system. Such a closed-loop operation makes it even more robust, in particular with respect to its own prediction errors. The method is therefore expected to be applied to a large variety of control means, in addition to the particular generation shedding considered in this chapter.

Finally, let us mention that the generation shedding scheme has shown to effectively apply to the two real-world power systems simulated so far, in addition to that considered in this chapter, namely: the Hydro-Québec system (in the Churchill Falls corridor) [Zhang et al., 1997b] and the Brazilian power system (Itaipu's site) [Ernst et al., 1998b].

5.2 Topics for further research work

Admittedly, the approach needs further developments and refinements. For example, the prediction scheme should be further validated on various types of power systems and, if necessary, modified accordingly.

The developments of this chapter have assumed that the contingency occurence and its clearance were properly identified. A systematic method should be set up to this end.

Another question is the real-time modeling of external equivalents. In the general case, we believe that current practices may readily suggest adequate solutions.

An important concern is how to appraise various types of control actions. The generation shedding considered in this paper was just a first attempt, and probably one of the less problematic to tackle.

Yet, another question worth developing concerns "local" control, where real-time emergency control would rely on local measurements relative to particular power plants, as opposed to the "global", centralized approach proposed here. This local approach would be less general and accurate than the global one, but at the same time less demanding in terms of information needed and corrresponding communication requirements.

The above list of open questions is certainly not exhaustive. But the main objective of this chapter has been to show that the general methodology is able to cope with the difficult problem of transient stability closed-loop emergency control and to pave the way towards pragmatic solutions.

6. SUMMARY

This chapter has described the fundamentals of a closed-loop emergency control scheme. It was developed in the particular case of generation shedding, and illustrated on a realistic situation.

The essential ingredients of such a scheme are: predictive TSA, predictive control, and real-time measurements: the first two tasks rely on SIME, the third on modern hardware and telecommunications facilities.

Unlike Preventive SIME which reached maturity and is now ready for use, Emergency SIME is still in its infancy. Nevertheless, it has promising skills, complementary to those of preventive control and also of open-loop emergency control.

Comparing the three types of control will be one of the objectives of the next and final chapter.

Chapter 7

RETROSPECT AND PROSPECT

This monograph has been devoted to a hybrid temporal-direct methodology, tailor-made for power systems transient stability assessment and control.
The objective of this final chapter is to give a general, quick overview by:

- *glancing over the monograph to summarize salient features of this general methodology as well as of its preventive and emergency versions*
- *giving an example to illustrate method's capabilities of extensions*
- *pointing out main differences and complementarities of this particular deterministic approach with the general statistical approach of automatic learning, as well as their ability to jointly cover a considerably broad field of applications.*

1. SIME: HINDSIGHT AND FORESIGHT
1.1 SIME: a unified comprehensive approach

SIME is a deterministic, tailor-made approach to power system transient stability.

It uses temporal information about the multimachine system continuously updated during the transients and transformed into that of an OMIB equivalent. The information is processed until meeting method's own instability/stability criteria.

The OMIB transformation reduces dramatically the dimensionality of the transient stability problem and provides three different representations of the phenomena, namely:

- OMIB swing curves: to identify the type of (in)stability modes and appraise impact of various system parameters

190 TRANSIENT STABILITY OF POWER SYSTEMS

- OMIB $P - \delta$ curves: to appraise margins and identify CMs
- OMIB phase plane: to get a synthetic view of transient stability phenomena.

These representations, together with the multimachine swing curves contain the necessary and sufficient information for deriving TSA&C techniques able to encounter needs of both preventive and emergency operation modes.

1.2 Preventive SIME

The preventive SIME gets information from contingency simulations performed step-by-step by a T-D transient stability program in order to set up a unified approach to transient stability

- analysis
- contingency screening
- sensitivity analysis
- control.

The tight coupling of the T-D program with the OMIB allows SIME to preserve the assets of these programs, namely

- accuracy
- flexibility with respect to power system modeling, contingency scenarios, modes of (in)stability
- compliance with existing operational strategies

and at the same time to speed up the assessment of T-D methods, while broadening enormously their possibilities. In particular, by

- complying with real-time requirements
- uncovering CMs and designing preventive control actions by quantifying judiciously generation shifting patterns
- providing transient stability-constrained ATC calculations and congestion management
- in short, by opening avenues to issues which have long been considered to be problematic, if at all feasible (e.g., see [Ilic et al., 1998]).

Further, interfacing SIME with an OPF program allows meeting in a near optimal way both static and transient stability constraints; it thus makes an old dream come true [Ribbens-Pavella et al., 1982a, Ribbens-Pavella et al., 1982b, Sterling et al., 1991]. Besides, this interface with OPF contributes to

Chapter 7 - RETROSPECT AND PROSPECT

cope with emerging market needs by helping the operator to make pertinent and transparent decisions.[1]

Finally, observe that the Preventive SIME is not a stand alone software; rather, it is coupled with the T-D program of concern.[2]

This coupling is a rather easy task.[3] We therefore believe that it is an important advantage of the method. Indeed, it offers the possibility of choosing the appropriate T-D programs, and hence the possibility for SIME to easily adjust itself to the modeling and specifics of the particular power system of concern. Over the years, SIME has thus been tested on a large variety of power systems, differing in their structure (radial vs meshed; e.g. the Hydro-Québec system vs the EDF power system [Zhang et al., 1997a]), by their size (e.g., the 3-machine system used for illustrations in this monograph vs the EPRI 627-machine test system) by the type of prevailing generation (hydro vs nuclear; others).

Throughout thousands of simulations performed on over 10 real-world power systems, SIME has behaved consistently well. The right choice of the T-D programs has contributed to its easy adaptation to the various power system.

An additional advantage of this coupling is that it makes SIME's interfacing with existing EMSs quite straightforward.

Note on instability phenomena and their characterization

Instability phenomena are certainly independent of the method used to analyze them.[4] However, SIME provides many interesting ways of characterizing and assessing them. For example, easy ways of appraising CCTs and PLs; or, for a given unstable contingency scenario, stability margins, number of involved CMs, time to instability, number of swings and their (negative) damping.

For illustration, let us consider again the stability case of the Hydro-Québec system described in § 4.2 of Chapter 2 and displayed on Figs 2.9. On this ex-

---

[1] Observe that other, more recent attempts to embed transient stability constraints in OPF algorithms, have also been proposed. For example in [De Tuglie et al., 1999] the idea is to combine transient stability and the conventional OPF problem into a single non-linear programming problem.
This type of approaches, however, lack transparency. Besides, they face the curse of dimensionality and cannot be directly solved by existing non-linear techniques within reasonable times. As an example, the model for a one-machine infinite bus test system represented by the conventional simplified model has 2002 variables and 3005 algebraic constraints for only 10 s simulation having a 10 ms step size length. A similar approach was also proposed by [Gan et al., 1998].

[2] For instance, the examples described throughout the monograph have used couplings with MATLAB (Section 1 of Appendix B), ST-600 (Sections 2 and 4 of Appendix B), ETMSP (Section 3 of Appendix B). Coupling with home-made T-D programs as well as with EUROSTAG and SIMPOW have also been mentioned in various publications (e.g., see [Zhang et al., 1997a, Ghandhari et al., 2000]).

[3] To fix ideas, a close collaboration of SIME's and T-D program's designers helps completing the job within less than one month. Subsequently, this job becomes rather trivial and takes about one week.

[4] Recall that one may distinguish many types of instabilities, such as first-swing vs multiswing, upswing vs backswings, plant vs inter-area modes ones and their combinations.

ample, one can observe the well-known fact that, in the presence of multiswing instability phenomena

- the first-swing CCT is larger than the multiswing one (here: 220 vs 175 ms)
- the time to instability is much smaller for first-swing than for multiswing (here: 0.418 vs 5.6 s).

To state this latter consideration otherwise, if the contingency is cleared at $220\,\text{ms} > t_e > 175\,\text{ms}$ the system will lose synchronism after some seconds (about 5 s), after a number of oscillations, while if $t_e > 220\,\text{ms}$ it will lose synchronism after some hundred ms (418 or less).

Notice also how clearly the number of swings and their damping are identified by SIME.

To illustrate inter-area instability modes, consider the case described in § 4.3.4 of Chapter 4, displayed on Figs 4.10, where 38 CMs swing against the remaining 50 machines. The example suggests that the CCT and time to instability of such phenomena may be quite small and close to each other. Along the same lines, for the stability case displayed on Figs 5.11, CCT = 156 ms ; if the contingency is cleared at $t_e = 167\,\text{ms}$, the system will lose synchronism at $t_u = 550\,\text{ms}$, i.e., less than 400 ms later.

1.3 Emergency SIME

The emergency SIME uses real-time measurements which warn about (i.e., detect) a contingency's actual inception (without necessarily identifying it). Processing these data allows SIME to set up a closed-loop emergency control which

(i) predicts the system instability ahead of time and assesses its characteristics (size, i.e. margin of the instability, CMs, time to instability)
(ii) devises an appropriate control action and sends the order of its triggering
(iii) continues monitoring the controlled system.

The three main ingredients of this closed-loop emergency control (CLEC) scheme are

1. predictive stability assessment
2. control
3. hardware for transferring information from power plants to control room and from this room to the controlled device (e.g., controlled plant(s)).

The speed (and hence duration) of a complete CLEC cycle is directly related to the possibilities of the telecommunications systems (hardware). Typically, the total duration is lowerbounded by about 350 ms but could be larger (450 or 500 ms). Recall, however, that the severer the instability and the shorter

Chapter 7 - RETROSPECT AND PROSPECT 193

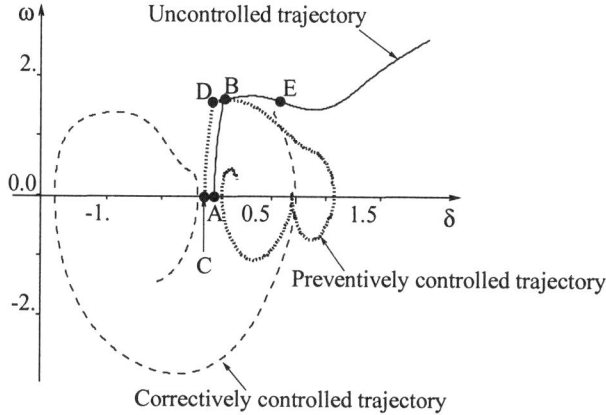

Notation:

A : original pre-fault stable equilibrium point
B : fault clearance of the uncontrolled trajectory
C : controlled pre-fault stable equilibrium point (preventive control)
D : fault clearance of the (preventively controlled) trajectory
E : application of the emergency control (shedding of 3 machines)

Figure 7.1. Difference between preventive and emergency control. Illustration on the OMIB phase plane computed for the 88-machine EPRI test system.

the duration required. Nevertheless, the CLEC scheme could be helpless for some extremely fragilized situations. In such cases, one should call upon either open-loop emergency control (OLEC) discussed in § 1.5, or on a combination of preventive and corrective controls, discussed below.

1.4 Preventive vs emergency control

Let us compare preventive control with emergency (or corrective) control, no matter whether it is of the open-loop or the closed-loop type.

1.4.1 Description of physical phenomena

While both approaches pursue the same objective, they use sort of complementary ways to reach it. Indeed, their common objective is to avoid loss of synchronism; in other words, to maintain the dynamic trajectory caused by a contingency inception within the domain of attraction of the system in its post-fault configuration. But to this end, preventive control acts also on the pre-fault operating conditions, while emergency control acts on the trajectory, in order to bend it.

Figure 7.1 describes the two procedures. It was obtained by simulations performed on the example of Chapter 6, using the OMIB phase plane, with the following notation:

- solid line: uncontrolled, unstable trajectory
- dotted line: controlled trajectory in the preventive mode, by means of a pre-contingency generation shifting of 589 MW
- dashed line: controlled trajectory in the emergency mode, by means of a post-contingency generation shedding of 2,240 MW, at $t_{ct} = 450$ ms after the fault inception.

Observe how clearly this phase plane representation describes the physical phenomena. These descriptions are complemented by the corresponding OMIB swing curves of Figs 7.2, which shed more light into the shapes of these phase portraits.

1.4.2 Controlled generation vs control time

Another interesting information is provided by the curve of Fig. 7.3. It was obtained via simulations carried out on the EPRI test system C. It displays the quantity of controlled generation ΔP_C necessary to preserve system's integrity vs the "control time", t_{ct}, (duration elapsed between the contingency inception and the actual application of the control action). Incidentally, observe that the first part of this curve is (almost) piece-wise linear. The preventive action is supposed to correspond to $t_{ct} \leq 0$. Besides, recall that this preventive action implies generation shifting, mainly from CMs to NMs. On the other hand, corrective action (be it closed-loop or open-loop) implies generation shedding.

1.4.3 Discussion

In terms of MW (amount of generation power), it is obvious that preventive control is much less demanding than emergency control. Indeed, it requires shifting rather than shedding generation; besides, the amount of shifted generation is smaller than the amount of generation shed. (In the previous example, 589 vs 2,240 MW.)

However, in terms of cost the comparison is less easy. Indeed, preventive control is based on hypothetical (though plausible) assumptions about anticipated operating conditions and contingency scenarios; but such conditions might never occur: the corresponding countermeasures become useless and the related cost unacceptably high. In contrast, emergency control actions are fully justified; the question is whether they could be triggered early enough to safeguard system's integrity.

The above discussion suggests that there could be a tradeoff between the size of countermeasures taken preventively and those taken correctively. For example, shifting preventively part of the required generation power (i.e., stabilizing only partly a contingency) might provide a better security vs cost solution: at a reasonable "cost", it would render the contingency less severe, resulting in a quite important saving of generation shedding in the emergency mode, when-

Chapter 7 - RETROSPECT AND PROSPECT

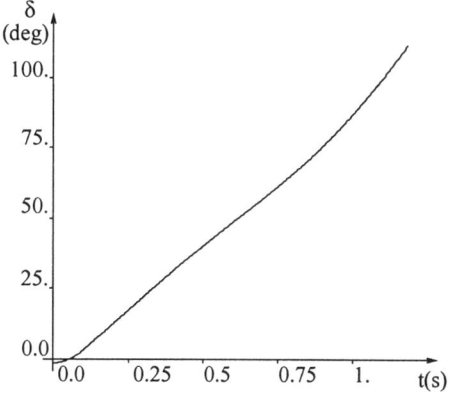

(a) Uncontrolled OMIB swing curve. Pre-fault: $P_C = 24,623$ MW

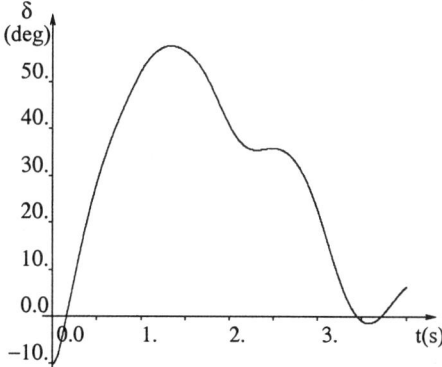

(b) Controlled OMIB swing curve in the preventive mode
Pre-fault: $P_C = 24,034$ MW

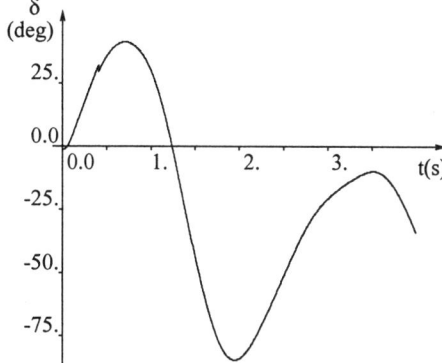

(c) Controlled OMIB swing curve in the corrective (emergency) mode
Pre-fault: $P_C = 24,623$ MW ; Post-fault: $P_C = 22,383$ MW

Figure 7.2. OMIB swing curves corresponding to the OMIB phase plane trajectories of Fig. 7.1

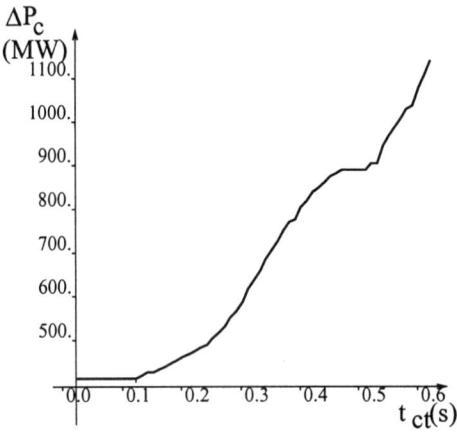

Figure 7.3. Quantity of controlled generation, ΔP_C, vs control time, t_{ct}. Adapted from [Ernst et al., 2000a]

ever this latter is needed, inasmuch as the resulting time to instability would increase quite a lot.

Incidentally, observe that the tools provided by SIME make the study of this kind of tradeoffs easy, systematic, and able to furnish near optimal solutions. This is further illustrated below.

1.5   General comparisons
1.5.1   Description

The real-time closed-loop emergency control (CLEC) proposed in Chapter 6, aims at replacing or complementing existing open-loop emergency control (OLEC) schemes. On the other hand, OLEC can be viewed as being half-way between CLEC and preventive control.

Let us first summarize the basics of these three types of control before comparing them further.

Preventive control aims at reinforcing the robustness of the (forecasted) system vis-à-vis (anticipated) plausible contingencies. The procedure relies on contingency simulations, performed off-line, in a horizon of, say, 30 min ahead.

Open-loop emergency control relies on off-line plausible (anticipated) contingency simulations to assess appropriate control actions, and on real-time measurements to detect anomalies and decide to trigger these actions.

Closed-loop emergency control relies on real-time measurements to predict in real-time (the size of) system's instability, and corresponding control actions, and to decide to trigger these actions.

Roughly, CLEC is composed of the steps (i) to (iii) mentioned in § 1.3, while OLEC uses part of step (ii).

Given today's hardware possibilities, OLEC is thus able to "save" 200-300 ms.

1.5.2 Preventive vs emergency control

In terms of amount of generation power, it is obvious that *under identical stability conditions* (contingency scenario and operating state), preventive control will be significantly less demanding than OLEC which will be less demanding than CLEC. Indeed, the earlier the control action is triggered, the earlier the predicted instability is contained and the smaller the size of this control. The curve of Fig. 7.3 illustrates this consideration on a concrete practical case.

However, the above assumption concerning application of various controls under identical stability conditions is quite unrealistic.

Indeed, as concerning preventive control, it is very unlikely that the forecasted stability conditions will actually occur, especially because of the combinatorial nature of the events. Hence, preventive control generally faces the tradeoff between security and economy in a sub-optimal way. Besides, today where new legislations open electricity markets, the trend is to avoid anti-economic stability countermeasures to be taken preventively; rather, they tend to operate the power systems increasingly closer to their stability limits: emergency control becomes the last resort for safeguarding system's integrity.

A midway solution suggested in § 1.4.3 would consist in applying preventively part of the countermeasure required, so as to mitigate the severity of the situation that the emergency task would have to control.

Open-loop vs closed-loop emergency control aspects are discussed below.

1.5.3 Open-loop vs closed-loop EC

Comparing OLEC with CLEC, we point out the following main differences.

Open-loop emergency control:

- is generally faster, and therefore cheaper; it can even be the only control action able to prevent loss of synchronism in some fragilized situations
- generally requires local measurements; it is therefore less expensive and less demanding in terms of communications hardware
- is spoilt by uncertainties: power system modeling uncertainties introduced during the off-line simulations (like preventive control); additional uncertainties concern the stability conditions for which the control actions are designed off-line
- is thus contingency dependent, i.e. pre-designed to cope with only contingencies very likely to arise

- once triggered, it has no means to re-adjust control actions, (open-loop technique).

Closed-loop emergency control:

- is slower and therefore more expensive
- requires more measurements than OLEC and is more demanding in terms of communications hardware
- is free from the above-considered uncertainties, and hence can assess (near) optimal control (and hence cheaper from this viewpoint) provided that the measurements are good enough (otherwise they replace one by another source of uncertainties)
- is thus able to cope with the actual contingency without necessarily knowing it, since the measurements take it into account
- being by essence a closed-loop technique, it continues monitoring and possibly controlling further the system, as long as needed.

All in all, OLEC and CLEC have complementary possibilities and can be used in combination, if deemed necessary.

2. AN ILLUSTRATION

This section illustrates on an example SIME's possibilities for further explorations of transient stability issues.

2.1 Description

In this paragraph, SIME is used to appraize the impact of the length of important transmission lines on transient stability. The study will be conducted using unusual events; but the obtained results are physically fully sound.

Contingencies considered so far were supposed to have a finite during-fault period (t_e) followed by a post-fault one. Generally, the post-fault configuration differs from the pre-fault one by the loss of transmission elements. We will suppose here that the duration of the fault is so small that it can be neglected ($t_e \to 0$) : the system stability is then determined by the transition from the initial to the final configuration.

Three situations may be distinguished, depending upon the system operating state in its final configuration.

1. No operating state exists anymore: this corresponds to the absence of solutions of the dynamic eq. (2.2), i.e., of equilibrium points in the system post-fault configuration.
2. A new operating state exists but the initial operating state does not belong to its domain of attraction.

3. A new equilibrium state exists and the initial operating state belongs to its domain of attraction.

2.2 Application

Below we focus on cases 2 and 3, that we illustrate on an example carried out on the EPRI test system C (see Section 3 of Appendix B). The final configuration differs from the initial one by the loss of two 500 kV identical parallel lines. They are located at a meshed part of the system and have an approximate length of 300 km. The power initially transferred through each line is 1,655 MW.

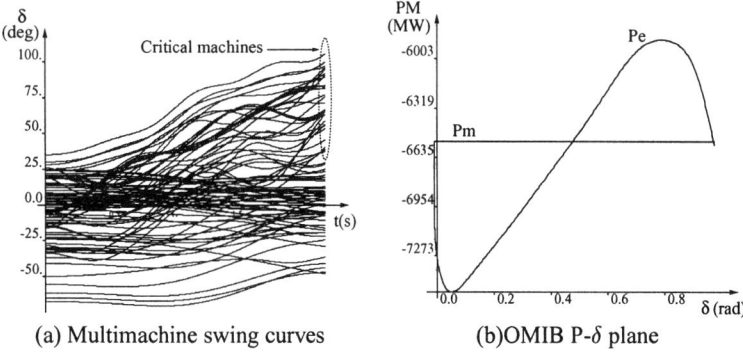

(a) Multimachine swing curves   (b) OMIB P-δ plane

Figure 7.4. Two lines tripped on the EPRI test system C

Figure 7.4a represents the swing curves of the multimachine system, corresponding to the case where both lines are opened at $t = 0$: 1.63s later the system loses synchronism.

This is corroborated by the OMIB transformation. The critical group is composed of 34 machines. Figure 7.4b displays the corresponding OMIB $P - \delta$ curves. At $t = 0$, its angle equals -0.011 rad. At $t = 0.92$ s ($\delta = 0.477$ rad), the OMIB accelerating power vanishes. The decelerating area that follows is not sufficient to bring the system back to synchronism and the unstable angle ($\delta_u = 0.974$ rad) is reached at $t_u = 1.63$ s.

Thus, although a post-fault equilibrium point exists, the system is unstable.[5]

2.3 Simulations results

Let us now assess the way the lines' length affects stability. To this end, let the two lines be connected to buses i and j, and express their parameters in

---
[5]Note that if only one line is tripped, the T-D simulation as well as SIME show that the system is stable. Note also that a steady-state simulator, which would consider the temporal evolution of the system as a succession of steady state equilibrium points, would not have discovered any instability.

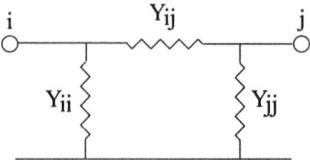

Figure 7.5. π modeling of the transmission elements

the three terms Y_{ii}, Y_{jj} and Y_{ij} of their π equivalent as (see Fig. 7.5):

$$Y_{ii} = Y_{ii}^{line} + Y_{ii}^{other} \ , \ Y_{jj} = Y_{jj}^{line} + Y_{jj}^{other} \text{ and } Y_{ij} = Y_{ij}^{line} + Y_{jj}^{other}$$

where superscript *line* denotes the participation of the two lines to the admittance matrix and *other* the participation of the other elements.

In the post-fault configuration these three terms can be respectively written as:

$$\beta * Y_{ii}^{line} + Y_{ii}^{other} \ , \ \beta * Y_{jj}^{line} + Y_{jj}^{other} \ , \ \beta * Y_{ij}^{line} + Y_{ii}^{other}$$

where

$\beta = 1$ means that the post-fault configuration is identical to the prefault one (two lines in service);

$\beta = 0.5$ or 0 means that in the post-fault configuration there are, respectively, one or zero lines in service.

More generally, β represents the fraction of the two lines in service.

Let us compute stability margins for values of $\beta \in [0, 1]$. Table 7.1 gathers the obtained results. The first column indicates the value of β, the second the unstable angle δ_u (the return angle δ_r if the system is stable), the third the stability margin and the last one the time to instability, t_u, (or the return time t_r if the system is stable). Observe that the stability margin increases with β; it becomes positive for β greater than 0.039: the system becomes stable.

Table 7.1. OMIB parameters for different values of β

β	$\delta_u(\delta_r)$ (rad)	η (MWs^2)	$t_{u(r)}$ (s)
0.000	0.976	-61.170	1.655
0.007	0.835	-30.435	1.660
0.015	0.718	-14.156	1.685
0.023	0.642	-7.988	1.710
0.031	0.574	-4.297	1.770
0.039	(0.550)	(> 0)	(2.255)

Figure 7.6 illustrates in the $P - \delta$ plane the OMIB P_e and P_m curves for the different values of β used in Table 7.1 (between brackets). Note that for $\beta = 0$, the P_e curve is reduced to a single point: the point where the other P_e curves take their origin.

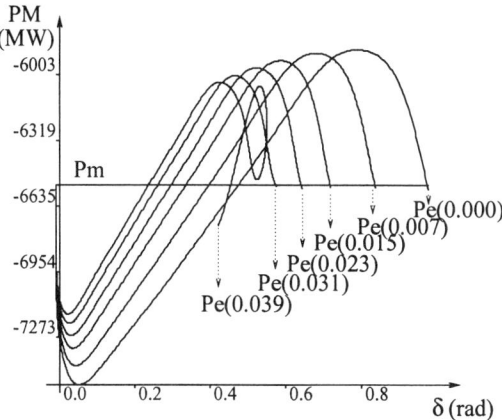

Figure 7.6. P_e for different values of β

The above SIME-based computations appraise the sensitivity of the system stability to the length of these heavily loaded transmission lines. The conclusions are sound from a physical point of view and interesting for practical applications.

3. COMPARING CLASSES OF METHODS

In Chapter 1 we have distinguished three classes of approaches to transient stability: the conventional T-D methods, the direct (pure or hybridized) methods and the automatic learning (AL) ones. The first two are deterministic approaches, the third is a statistical one.

This monograph has been devoted to SIME, a particular hybrid temporal-direct method, dedicated to transient stability. On the other hand, T-D, and to a larger extent AL methods can cover a much broader field of applications. However, in this section the comparisons will concentrate on transient stability. They are summarized in Table 7.2.

Note that while SIME – and as a corollary T-D methods – have received extensive consideration, AL methods have been described only very shortly in Section 6 of Chapter 1. Below we give a brief complement to this "digest" before proceeding with general comparisons.

Table 7.2. Transient stability assessement: techniques vs criteria. Adapted from [CIGRE, 1997]

Criteria:	1	2	3	4	5	6	7	8	9	10
		Information required		Off-line	Real-time	Information provided				
Method	Model. poss.	Off-line	Real-time	prep. tasks	comput. reqts	Anal.	Screen.	Inter-pret.	Sensit.	Control	Managt uncert.
T-D	+++	FD	CS	mV	B	D	–	–	–	–	
AL											
- Off-line prepar.	+++	FD		DB,V		D,S		+	+		+
- Real-time use	+++		SV		EL	D,S		+	+	P,C	+
SIME											
- Preventive	+++	FD	CS	mV	FB	D,S	++	+	+	P	–
- Emergency	+++	–	SV	mV	EL	S		–	+	C	+

3.1 Practical aspects of AL approaches

3.1.1 Real-world applicability concerns

The short description in Section 6 of Chapter 1 suggests that AL methods can meet stringent needs of power-system security. The practical application reported [Cholley et al, 1998] is just an example of the wide range of the possibilities they offer.

A prerequisite to the implementation and use of AL methods is the systematic data base generation. In addition, in order for AL methods to be accepted by the power system community, new strategies should be devised, generally different from those developed in the context of deterministic approaches. Thus, AL methods are likely to be used at first in areas where they are able to supplement deterministic approaches. Such an important contribution could be made by an approach able to handle simultaneously many security aspects [Wehenkel et al., 1997].

Maintaining expertise in off-line study environments is still another way of taking due advantage of statistical approaches.

3.1.2 SIME as compared with AL approaches

While SIME is a technique dedicated to TSA&C, the scope of AL methods is significantly broader, with extraordinarily rich and diversified practical outcomes. In particular, AL methods can cover the entire field of security. Along these lines, one major contribution of AL methods could precisely be a unified approach to power system security [Wehenkel, 1999]. This, of course, implies a heavy panoply of data mining techniques and a close collaboration between designers of such techniques with power engineers.

In the field of TSA&C, SIME and AL approaches have complementary features. Incidentally, SIME may preanalyze stability scenarios (for example in terms of stability limits, stability margins, critical machines, sensitivity analysis techniques) for use in the data base, from which synthetic information is extracted in terms of decision trees, ANNs and the like.

3.2 Synthetic comparison

In this paragraph a synthetic comparison is given of the three classes of transient stability methods: T-D vs SIME vs probabilistic automatic learning approaches. Table 7.2 displays the comparisons. They are carried out in terms of the various criteria described below.

Obviously, such a synthetic comparison is inevitably superficial; it only intends to point out general trends.

3.3 The criteria

Column 1. *Modeling possibilities*: ability of the method to embed more or less detailed models

+++ : Fully detailed power system modeling.

Column 2. *Type of information required* (with respect to the most detailed possible modeling):

- Off-line:

    FD : Fully Detailed information, including controllers

    PD : Partly Detailed information, concerning essential parameters relevant to the considered security problem

    LF : Load Flow type of information

- Real-time:

    CS : Complete System picture, e.g. as provided by state estimation

    SV : Selected Variable (a very few number of electrical and topological variables), directly measured or easily computed, e.g. as provided by SCADA.

Column 3. *Off-line preparation tasks*, mainly for the purpose of:

- Validating and building a data base
- Validation, including evaluation of the applicability domain, tuning of parameters, etc., is usually carried out every time a method is applied to a new power system or when significant changes appear in the system. One

distinguishes tasks requiring:

V : Validation and,

mV : minor Validation,

DB : Data Bases, on the other had, are built for the purpose of intelligent systems, using human expertise and/or other methods, of the system theory type; a data base has to be refreshed every time the system configuration and operating conditions change "significantly". Generally, this is required much more often than validation, and hence implies more massive computations.

Column 4. *real-time computational requirements*

B : Bulky computations, e.g. required for time-domain simulations with full model,

FB : Fairly Bulky computations, e.g. reguired for time-domain simulations with simplified model,

L : Light computations, e.g. required for fast direct methods,

EL : Extremely Light computations, e.g. required when using a decision tree to assess the stability of a state.

To fix ideas, B may correspond to hours of CPU time, FB to minutes, L to seconds and EL to fraction of millisecond CPU time.

Column 5. *Type of information provided for analysis:*

D : Detailed (e.g. bus voltages, swing curves, etc.); the type of provided detailed information may change from one method to the other.

S : synthetic (e.g. margins).

Column 6. *Contingency screening tool.*

A fast method may be not accurate enough (e.g. when complying with simplified system modeling only), yet reliable enough to identify and filter out "uninteresting contingencies", i.e. generally contingencies which are too mild to be dangerous. In general, the more simplified the system model, the simpler and faster the method.

++ : Ultra-fast screening tool.

+ : Very fast screening tool.

Column 7. *Interpretability of phenomena:* method's ability to uncover and describe the main mechanism driving the phenomena of concern:

+ : very transparent information ,

− : "black-box" type of information (e.g., yes or no answer about stability; except for time-domain response).

Column 8. *Sensitivity analysis:* in terms of computational overhead.

+ : easy (e.g. via extrapolation-interpolation of margins vs parameters),
− : difficult (cut-an-try, e.g. dichotomic search).

Column 9. *Means to control:*
P : Preventive, i.e. means to reinforce system robustness, C : Corrective, i.e. means to bring the system back from emergency to normal.

Column 10. *Management of uncertainties*: ability to account for uncertainties inherent to modeling imperfections, partly/momentarily missing data, unknown data, etc.
+ : ability
− : inability.

3.4 Comments

1. Methods belonging to the same class have not necessarily equivalent performances; the table takes into account the best of the existing performances. Indeed, the table attempts to highlight strengths of a class of methods, even if some of its members do not meet them. E.g., only some automatic learning methods have the ability of interpretability. On the other hand, interpretability may have various meanings; e.g. a decision tree describes the physical phenomena at hand in a synthetic way (the parameters and their threshold values driving the phenomena), whereas a k-NN uncovers similarities of a given state with its neighbour states.

2. The table highlights complementary rather than competitive aspects of the various techniques.

3. In terms of computational performances, the effect of power system size is not considered here. This depends very strongly on various factors, in particular method implementation, software and hardware environments, etc. Similarly, the possibility of exploiting parallel computations is not taken into account.

4. An empty box means that the considered criterion is not of concern for the corresponding technique.

4. POSTFACE

Over the four past decades, the necessity of implementing in the control center dynamic security assessment functions has been repeatedly claimed; without success: adequate techniques were missing and the necessity of their use was not felt strongly enough.

After this 40-year (seeming) lethargy, we start now moving fairly fast: from lack of techniques to a panoply of techniques with overlapping possibilities. And demanding practicing engineers replace reluctant ones.

A key issue for good choices and adequate solutions is the synergy of practicing engineers, researchers and developers.

Appendix A
THE EQUAL-AREA CRITERION

1. GENERAL CONCEPTS
1.1 Introduction

The equal-area criterion (EAC) is an old graphical method that allows assessing the transient stability of electric power systems in a simple and comprehensive way. This method was developed and popularized at the end of the 30's and its origin is not very well known. Most of the first references to EAC are often made in books like [Dahl, 1938, Skilling and Yamakawa, 1940, Kimbark, 1948], which are among the first to describe and use it, mainly for the assessment of transient stability of one-machine connected to an "infinite" bus[1] (or of a two-machine system).

One of the main appealing characteristics of EAC is that its use eliminates the need of computing the swing curves of the system, thus saving a considerable amount of work, even if, in its "pure" statement, very simplified assumptions were made regarding power system modeling. Indeed, the system was represented by the classical model having the following features[2]:

- synchronous machines are represented by a constant voltage source behind the transient reactance
- synchronous machines have constant mechanical power and negligible damping
- loads are represented by constant impedance characteristics.

---
[1] The infinite bus being referred is regarded as a synchronous machine having zero impedance and infinite inertia, not affected by the amount of current drawn from it. It is a source of constant voltage (both in phase and magnitude) and frequency.
[2] Classical model, that is considered still valid for assessing first swing stability, is referred in this monograph as "Simplified Modeling" (SM) and used mainly in FILTRA for contingency filtering purposes. (See Chapter 5.)

Later, it was shown in the different applications and extensions of the method (ending with SIME method), that these simplifying assumptions in system modeling were not necessary for the method to work, and that it is also useful for assessing stability phenomena more complicated than first swing instabilities like multiswing and backswing ones, in large multimachine systems.

As described below, EAC relies in the concept of energy and is a powerful tool for assessing stability margins and limits, and for appraising the influence on stability of various system parameters. It is also very useful for explaining preventive and emergency control actions taken in industry for improving stability and, in general, for providing insight into the very physical phenomena.

1.2 Principle

Consider a machine connected to an infinite bus system.

$$M\frac{d^2\delta}{dt^2} = P_m - P_e = P_a \tag{A.1}$$

where

M = inertia coefficient of the machine
δ = angle between the machine's rotor and the infinite bus
P_m = mechanical power input to the machine
P_e = electrical power output of the machine
P_a = accelerating power.

From eq. (A.1)

$$\frac{d\delta}{dt} = \omega - \omega_0 \tag{A.2}$$

$$M\frac{d\omega}{dt} = P_m - P_e = P_a \tag{A.3}$$

where

ω = rotor speed of the machine
ω_0 = rated angular frequency of a synchronously rotating frame.

Multiplying both sides of eq. (A.1) by $\frac{d\delta}{dt}$ results in

$$M\frac{d^2\delta}{dt^2}\frac{d\delta}{dt} = P_a\frac{d\delta}{dt} \tag{A.4}$$

Hence

$$\frac{M}{2}\frac{d[(\frac{d\delta}{d2})^2]}{dt} = P_a\frac{d\delta}{dt} \tag{A.5}$$

Multiplying both sides of eq. (A.5) by dt to obtain differentials instead of derivatives, we have

$$\frac{M}{2} d\left[\left(\frac{d\delta}{dt}\right)^2\right] = P_a d\delta. \qquad (A.6)$$

Integrating the equation between the pre-fault stable equilibrium angle δ_0 to any (during- or post-fault) angle δ yields

$$\frac{M}{2}\left(\frac{d\delta}{dt}\right)^2 = \int_{\delta_0}^{\delta} P_a d\delta \qquad (A.7)$$

or, by virtue of eq. (A.2),

$$\frac{M}{2}(\omega - \omega_0)^2 = \int_{\delta_0}^{\delta} P_a d\delta. \qquad (A.8)$$

Hence

$$\omega - \omega_0 = \sqrt{\frac{2}{M}\int_{\delta_0}^{\delta} P_a d\delta}. \qquad (A.9)$$

In the above equation, $(\omega - \omega_0)$ is the relative speed of the machine with respect to the infinite bus. If the system is first swing stable then this speed must return to zero when the accelerating power is either zero or of opposite sign to the rotor speed. For example, a monotonously increasing rotor angle δ implies $\omega - \omega_0 > 0$; the angle δ stops increasing at a maximum value δ_m when $\omega - \omega_0 = 0$; this only happens when a negative accelerating power P_a damps the speed from $\omega - \omega_0 > 0$ to 0. This process can be expressed as follows:

$$\omega - \omega_0 = \int_{\delta_0}^{\delta} P_a d\delta > 0 \quad \delta_0 < \delta < \delta_m; \qquad (A.10)$$

$$\omega - \omega_0 = \int_{\delta_0}^{\delta_m} P_a d\delta = 0 \quad P_a(\delta_m) \leq 0. \qquad (A.11)$$

If the accelerating power is plotted as a function of δ as shown in Fig. A.1, eqs (A.10) and (A.11) can be interpreted as composed of two areas, one accelerating (positive) area A_{acc} and one decelerating (negative) area A_{dec}; these two areas become equal for $\delta = \delta_m$ with $P_a(\delta_m) \leq 0$. The stability limit occurs when $A_{acc} = A_{dec}$ and $P_a(\delta_m) = 0$; the system is then said to be *critically stable*. This is the definition of the EAC, stated in the following equation:

$$P_a(\delta_m) \leq 0, \text{ and } \int_{\delta_0}^{\delta_m} P_a d\delta = \int_{\delta_0}^{\delta_m} (P_m - P_e) d\delta = 0. \qquad (A.12)$$

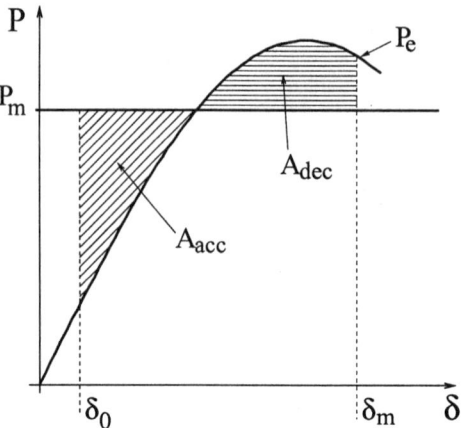

Figure A.1. Equal-area criterion

For this condition, δ_m coincides with the unstable equilibrium point $(\pi-\delta_P)$ with δ_P as the post-fault steady-state stable equilibrium angle for the OMIB system shown in Fig. A.2. If the pre-fault and post-fault configurations are the same, then δ_m coincides with the unstable equilibrium point $(\pi - \delta_0)$ shown in Fig. A.2.

Let us now assume that the machine is connected to an infinite bus by a double circuit transmission line. Further, let us assume that the fault occurs on one of the lines, and search for the critical clearing angle, δ_c. Figure A.2 shows the power-angle curves for the three different cases, namely pre-, during- and post-fault: the power-angle curves are generally displaced sinusoids. To obtain the critical clearing angle we equate the two areas A_{acc} and A_{dec} which, according to Fig. A.2, are expressed as

$$A_{acc} = \int_{\delta_0}^{\delta_c} (P_m - P_{eD}(\delta))\, d\delta \qquad (A.13)$$

$$A_{dec} = \int_{\delta_c}^{(\pi-\delta_P)} (P_{eP}(\delta) - P_m)\, d\delta. \qquad (A.14)$$

Integration of $P_{eD}(\delta)$ and $P_{eP}(\delta)$ does not pose any problems. Thus for getting δ_c the condition is

$$\int_{\delta_0}^{\delta_c} (P_m - P_{eD}(\delta))\, d\delta = \int_{\delta_c}^{(\pi-\delta_P)} (P_{eP}(\delta) - P_m)\, d\delta. \qquad (A.15)$$

This involves the solution of a simple trigonometric equation.

Appendix A: THE EQUAL-AREA CRITERION 211

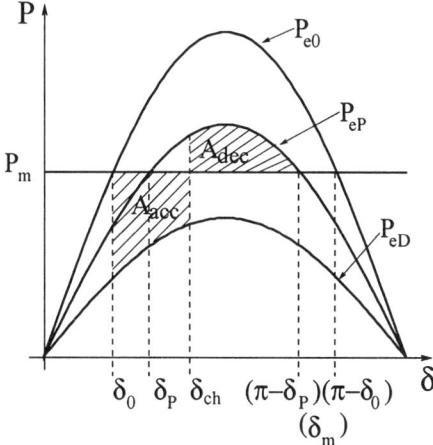

Figure A.2.   Equal-area criterion for obtaining the critical clearing angle.

1.3 Two-machine system

For applying the equal-area criterion to a two-machine system it is necessary to reduce this system to a one-machine infinite bus one. In this way, all the concepts developed above are applicable. The derivation of the swing equations of the equivalent system follows the pattern used in § 2.2 of Chapter 2.

2. APPLICATION EXAMPLE

The main objective of this section is to show the curves obtained when using this "pure" EAC together with a T-D pogram. To this end, we consider the one-machine infinite bus system displayed in Fig. A.3 and plot the EAC curves using a MATLAB. The system has one synchronous machine connected to an infinite bus through two identical parallel lines. The contingency consists of a three-phase short-circuit applied at the middle of one of the transmission lines (as shown in the figure), that is tripped when clearing the fault. The synchronous machine is represented using the classical model; its dynamic data and initial operating condition are displayed in the figure.

The resulting power-angle curves are displayed in Fig. A.4 for two different clearing times: one slightly larger, the other slightly smaller than CCT.

An interesting aspect of this application example is to stress that, when using the EAC together with a T-D program, it is not necessary to compute the areas under the electrical power curve because the system response naturally makes the electrical curve either to cross the mechanical power curve at point δ_u (unstable case) or to go back after reaching the angle δ_r at which A_{acc} and A_{dec} become equal (stable case).

212   TRANSIENT STABILITY OF POWER SYSTEMS

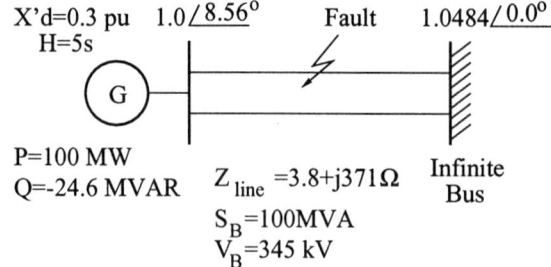

Figure A.3.   Data of the test system for the EAC application example

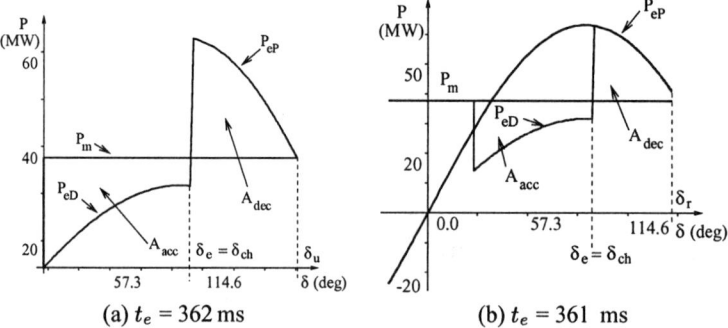

Figure A.4.   Power-Angle curves for the considered contingency under different clearing times.

This observation is readily generalized to the EAC application to a multimachine system after its reduction to the relevant OMIB. Expressions (2.19) and (2.21) calculate respectively the unstable and stable margins.

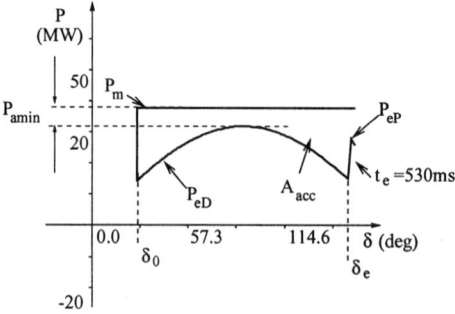

Figure A.5.   Power-Angle curve of a "too unstable" case.

Finally, let us consider a "too unstable" case, where the curves P_m and P_e do not cross anymore. This case is displayed in Fig. A.5. This case complements the discussion of § 3.3 of Chapter 2. We see that for making the system recover a post-fault equilibrium point, it is necessary to change either the mechanical power (if the clearing time is fixed) or the electrical power (if the mechanical power is fixed). The P_{amin} expressed by eq. (2.30) is clearly indicated.

Appendix B
DATA OF SIMULATED SYSTEMS

This appendix provides data on a sample of power systems with various

- *characteristics, in terms of: size (number of machines, buses, lines); prevailing type of generation; structure (radial vs meshed)*
- *models (of machines, loads and transmission network elements)*
- *types of transient (in)stability: first- or multi-swing; plant- or inter-area mode*
- *operational practices (T-D program; (in)stability criteria; maximum integration period)*
- *types of studies: preventive, for analysis, sensitivity analysis and preventive control; emergency, for predictive assessment and corrective control.*

The systems have been used throughout the monograph to illustrate SIME's specifics and performances. Detailed system data like machine and transmission network parameters are given below only for the small three-machine test power system described in Section 1.

1. THREE-MACHINE TEST SYSTEM

Due to its simplicity, this test power system is used for illustrating the basic principles of SIME and derived techniques for preventive TSA&C. It was adapted from [EPRI, 1977] and also used in [Sauer and Pai, 1998, Anderson and Fouad, 1993].

Figure B.1 displays its one-line diagram and Table B.1 its main characteristics. Table B.2 describes the dynamic data of the machines and their corresponding excitation control system. This latter is an IEEE Type 1 AVR, presented in Fig. B.2.

Simulation data of the contingencies used in the examples are shown in Table B.3.

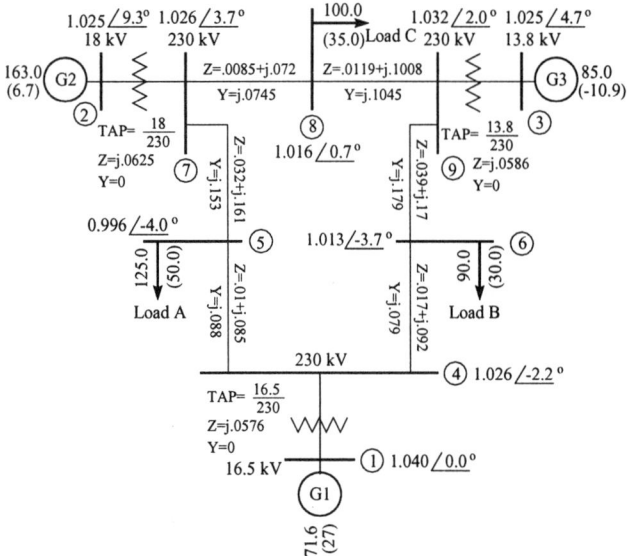

Figure B.1. Three-machine test power system. Adapted from [EPRI, 1977]

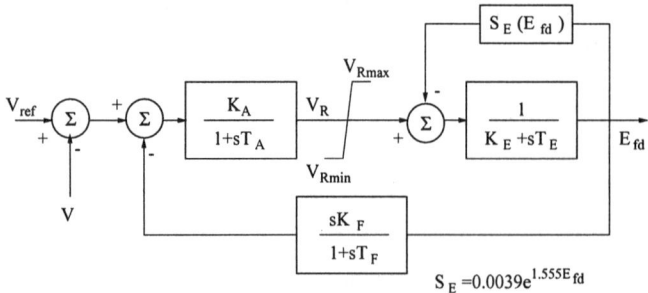

Figure B.2. Automatic Voltage Regulator Type I model. Adapted from [IEEE, 1968]

Table B.1. Three-machine test system main characteristics. Adapted from [EPRI, 1977]

Nr of buses	9
Nr of lines	9
Nr of detailed generators	3
Load model	Constant impedance characteristic

The simulations use SIME coupled with MATLAB [MATLAB, 1999].

On a stable T-D simulation, the maximum integration period is 3 s.

Table B.4 describes the contingencies considered for the simulations on the three-machine system. They all consist of applying a three-phase short-circuit to one bus and tripping one line when clearing the fault. Thus, two

Appendix B: DATA OF SIMULATED SYSTEMS   217

Table B.2.   Machine and exciter data for the three-machine test system

	Machine Data				Exciter Data *		
Parameters		m_1	m_2	m_3	Parameters		All machines
H	(s)	23.64	6.4	3.01	K_A		20
M	(s²/rad)	12.54	3.39	1.59			
X_d	(pu)	0.146	0.8958	1.3125	T_A	(s)	0.2
X_d'	(pu)	0.0608	0.1198	0.1813	K_E		1.0
X_q	(pu)	0.0969	0.8645	1.2578	T_e	(s)	0.314
X_q'	(pu)	0.0969	0.1969	0.25	K_F		0.063
T_{do}'	(s)	8.96	6.0	5.89	T_F	(s)	0.35
T_{qo}'	(s)	0.31	0.535	0.6			

* See Fig. B.2.

Table B.3.   Three-machine power system simulation data

Nr of pre-contingency operating states	1
Nr of contingencies	12
Type of contingencies	Three-phase short circuit applied at one bus and cleared by tripping one line
Event times	0-10 cycles
Simulation time	3 seconds

Table B.4.   Contingencies of the three-machine system

Cont. Nr (faulted bus)	Cont. Nr (faulted bus)	Line tripped
1 (5)	2 (7)	5 - 7
3 (7)	4 (8)	7 - 8
5 (8)	6 (9)	8 - 9
7 (9)	8 (6)	9 - 6
9 (6)	10 (4)	6 - 4
11 (4)	12 (5)	4 - 5

contingencies were defined for each of the lines shown in the third column. The values in the two first columns give the identification number and, between brackets, the faulted node for each contingency.

2.   HYDRO-QUEBEC POWER SYSTEM

Hydro-Québec power system main characteristics and simulation data are displayed in Tables B.5 and B.6 respectively. A very general one-line diagram showing the structure of the system is given in Fig. B.3.

The simulations are performed by SIME coupled with the Hydro-Québec transient stability program ST-600 [Valette et al., 1987].

On a stable simulation, the maximum T-D integration period is 10 s.

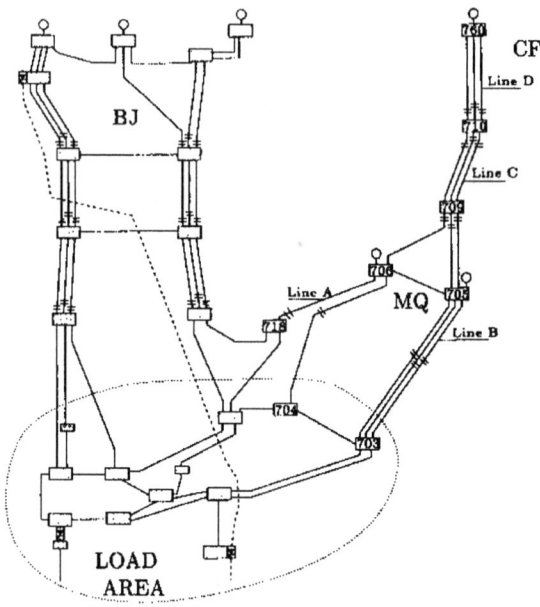

Figure B.3. One-line diagram of the Hydro-Québec power system. Adapted from [Zhang et al., 1996]

Table B.5. Hydro-Québec power system main characteristics

Nr of buses	661
Nr of lines	858
Nr of detailed generators	86
Nr of synchronous condensers	8
Load model	Nonlinear load characteristic
Total power	36,682 MW

Table B.6. Hydro-Québec power system simulation data

Nr of pre-contingency operating states	1
Nr of contingencies	377
Type of contingencies	Three-phase short-circuit applied at one bus and cleared by tripping one line
Event times	95 ms
Simulation time	10 seconds

3. EPRI AMERICAN TEST SYSTEMS

N.B. The data presented in this section are adapted from [EPRI, 1995]. These data are used by some utilities in North America and represent real electrical

Appendix B: DATA OF SIMULATED SYSTEMS 219

power systems. For confidentiality reasons, the original data provided by the utilities have been modified and the resulting data are used here only as a realistic test bed for the TSA&C techniques developed in this monograph.

The T-D program coupled with SIME for simulating these systems is ETMSP [EPRI, 1994].

3.1 Test power system C

Test power system C is the smallest test system given in [EPRI, 1995]. Its main features and simulation data are displayed in Tables B.7 and B.8.

On a stable simulation, the maximum T-D integration period is 5 s.

Table B.7. EPRI test system C main characteristics. Adapted from [EPRI, 1995]

Nr of buses	434
Nr of lines	2357
Nr of classical generators	74
Nr of detailed generators	14
Load model	Constant impedance characteristic
Total power	350,749 MW

Table B.8. EPRI test system C simulation data. Adapted from [EPRI, 1995]

Nr of pre-contingency operating states	7
Nr of contingencies	36
Pre-contingency/contingency combinations	252
Type of contingencies	Three-phase short circuit applied at one bus and cleared by tripping one or several lines
Event times	81.5 ms, 92 ms, 95 ms
Simulation time	5 seconds

3.2 Test power system A

Test power system A is the largest test system provided in [EPRI, 1995]. Its main features and simulation data are displayed in Tables B.9 and B.10. If two state variables are assigned to each classical synchronous machine model (301 machines) and a minimum of seven state variables to each detailed model (376 machines), the system is thus represented by over 3200 state variables.

Table B.10 shows that this system considers "remedial measures" for controlling some instabilities, making it very useful for comparing the effects of

Table B.9. EPRI test system A main characteristics. Adapted from [EPRI, 1995]

Nr of buses	4112
Nr of lines	6091
Nr of generators	**627** (346 with detailed and 281 with classical model)
Synchronous condensers	**36** (25 with detailed and 11 with classical model)
Load model	2240 Nonlinear static load characteristics
	14 large dynamic loads (synchronous motors, 5 with detailed and 9 with classical model)
Nr of SVC's	2
Total generation	76,1709 MW and 24,315 MVar

Table B.10. EPRI test system A simulation data. Adapted from [EPRI, 1995]

Nr of pre-contingency operating states	1
Nr of contingencies	22
Type of contingency	applying a three phase fault to one bus and tripping one or several lines when clearing the fault
Remedial Schemes	*Scheme 1:* at 0.25 s after the fault application, shed loads at buses 1219, 1351, 793, 1267, 1806, 842, 1826, 1757, 999 and 965 for a total amount of 2400 MW
	Scheme 2: Dropping pumps in the following steps (all times are measured after the fault inception):
	− at 0.2333 s, 370 MW in buses 2230 and 2411
	− at 0.25 s, 350 MW in buses 2617 and 2069
	− at 0.3 s, 96 MW of pumps (bus 2090) and 1432 MW of generation on the sending end (buses 421, 422, 3357 and 3237)
	Ramp up DC line (between buses 485 and 947) in 10 % steps at times: 0.4 s, 0.5 s, 0.6 s, 0.7 s, 0.8 s for a total of 1000 MW increase in DC flow
	Scheme 3: Dropping pumps in the following steps (all times are measured after the fualt inception):
	− at 0.2333 s, 300 MW in bus 2230
	− at 0.25 s, 177 MW in bus 2069
	− at 0.30 s, 96 MW in bus 2090
Event times	0-4 cycles (approx. 66 ms)
Maximum simulation time	10 seconds
Stable cases	21 (with respect to the event times)
Unstable cases	1 (with respect to the event times)

Appendix B: DATA OF SIMULATED SYSTEMS 221

remedial (or emergency) control actions with the ones of preventive control actions.

Original simulation data from [EPRI, 1995] was modified by creating additional contingencies from each one of the four contingencies which simulate remedial schemes (described in Table B.10). Thus, each one of contingencies 1, 4, 5 and 7 give rise to two contingencies:

- one which does not simulate the remedial scheme (Contingencies Nr 1, 4, 5 and 7)
- another contingency modeling the scheme (Contingencies Nr 23, 24, 25 and 26).

4. BRAZILIAN POWER SYSTEM

A model of the Brazilian South-Southeast-Centerwest power system described in [Bettiol, 1999] is considered in the present monograph. This model, which represents a major part of the Brazilian power system (approximately 80% of the electric energy market), is shown schematically in Fig. B.4. Main characteristics and simulation data of this test power system are displayed in Tables B.11 and B.12.

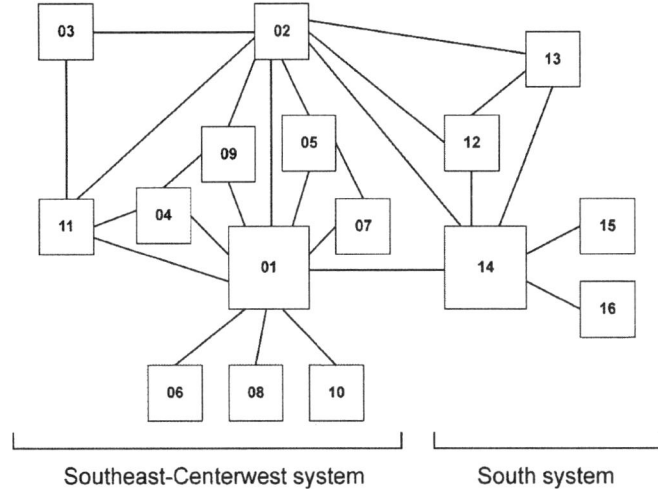

Figure B.4. Schematic diagram of the Brazilian South-Southeast-Centerwest power system. Adapted from [Bettiol, 1999]

The T-D program coupled here with SIME is ST-600 [Valette et al., 1987].

On a stable simulation, the T-D simulation period is 4 s when using the detailed model (DM).

The classical simplified model (SM) is available and valid for this power system. This advantage is used when implementing the Filtering Ranking

Table B.11. Brazilian South-Southeast-Centerwest system main characteristics. Adapted from [Bettiol, 1999]

Nr of buses	1185
Nr of lines	1990
Nr of detailed generators	56
Load model	Constant impedance characteristic
Total power	33,299 MW

Table B.12. Brazilian South-Southeast-Centerwest system simulation data. Adapted from [Bettiol, 1999]

Nr of pre-contingency operating states	2
Nr of contingencies	850
Pre-contingency/contingency combinations	1700
Type of contingencies	Three-phase short circuit applied at one bus and cleared by tripping one or several lines
Event times	0-10 cycles (approx 167 ms)
Simulation time	4 s for DM and 1 s for SM

and Assessment (FILTRA) scheme (see Section 2 of Chapter 5). The two pre-contingency operating states correspond to the plant- and inter-area mode examples given in Chapter 5.

References

[ABB, 2000] ABB Power Systems. "SIMPOW, Power System Simulation and Analysis Software: User's Manual". *ABB Power Systems Analysis Dept*, Vasteras, Sweden, 2000.

[Alaywan and Allen, 1998] Z. Alaywan and J. Allen. "California Electric Restructuring: A Broad Description of the Development of the California ISO". *IEEE Trans. on PWRS*, Vol. 13, No 4: 1445-1452, November 1998.

[Anderson and Fouad, 1993] P.M. Anderson and A.A. Fouad. *Power System Control and Stability*, Revised printing, IEEE Power System Engineering Series, IEEE Press Inc., 1993.

[Athay et al., 1979] T. Athay, V.R. Sherkat, R. Podmore, S. Virmani, and C. Pench. "Transient Energy Stability Analysis". *System Engineering for Power: Emergency Operating State Control - Section IV*. U.S. Department of Energy Publication, No. CONF-790904-P1, 1979.

[Avila-Rosales et al., 2000] R. Avila-Rosales, D. Ruiz-Vega, D. Ernst and M. Pavella. "On-Line Transient Stability Constrained ATC Calculations". *IEEE PES/Summer Meeting*, Seattle, July 16-20, 2000.

[Aylett, 1958] P.D. Aylett. "The Energy-Integral Criterion of Transient Stability Limits of Power Systems". In *IEE Proc.*, 105-C:527–536, 1958.

[Bellman, 1962] R. Bellman. "Vector Lyapunov Functions". *SIAM J. Control*, 1: 32, 1962.

[Bergen and Hill, 1981] A.R. Bergen, and D.J. Hill. "A Structure Preserving Model for Power System Stability Analysis". *IEEE Trans. on PAS*, PAS-100: 25–33, 1981.

[Bettiol et al., 1996] A.L. Bettiol, Y. Zhang L., Wehenkel, and M. Pavella. "Transient Stability Assessment of the South Brazilian Network by SIME

- A Preliminary Study". In *Proc. of North American Power Symposium '96*, MIT, Boston, MA, USA: 151-158, November 1996.

[Bettiol et al., 1997] A.L. Bettiol, Y. Zhang, L. Wehenkel, and M. Pavella. "Transient Stability Investigations on a Brazilian Network by SIME". In *Proc. of APSCOM '97*, Hong Kong, November 1997.

[Bettiol et al., 1998] A.L. Bettiol, L. Wehenkel, and M. Pavella. "Generation Allocation for Transient Stability-Constrained Maximum Power Transfer". In *Proc. of SEPOPE '98*, Salvador, Brazil, June 1998. (Invited paper, IP039; 11 pages).

[Bettiol, 1999] A.L. Bettiol. *Maximum Power Transfer in Transient Stability-Constrained Power Systems: Application to a Brazilian Power Network*. PhD Thesis, University of Liège, 1999.

[Bettiol et al., 1999a] A.L. Bettiol, L. Wehenkel, M. Pavella. "Transient Stability Constrained Maximum Allowable Transfer". *IEEE Trans. on PWRS*, Vol. 14, No 2: 654-659, May 1999.

[Bettiol et al., 1999b] A.L. Bettiol, D. Ruiz-Vega, D. Ernst, L. Wehenkel, and M. Pavella. "Transient Stability-Constrained Optimal Power Flow". In *Proc. of the IEEE Power Tech'99*, Budapest, Hungary, Aug. 29th - Sept. 2nd, 1999.

[Chiang et al., 1991] H.D. Chiang, F.F. Wu, and P.P. Waraiya. "A BCU Method for Direct Analysis of Power System Transient Stability". *IEEE/PES Summer Meeting*, San Diego, USA, Paper 91SM423-4 PWRS, July-August 1991.

[CIGRE, 1997] CIGRE Task Force 38.02.13. "New Trends and Requirements for Dynamic Security Assessment". Convener: B. Meyer. December 1997.

[Cholley et al, 1998] P. Cholley, C. Lebrevelec, S. Vilet and M. de Pasquale. "A Statistical Approach to Assess Voltage Stability Limits". In *Bulk Power System Dynamics and Control IV - Restructuring*, Santorini, Greece: 219-224, August, 1998.

[Christie et al., 2000] R.D. Christie, B.F. Wollemberg and I. Wangensteen. "Transmission Management in the Deregulated Environment". In *Proc. of the IEEE*, Vol. 88, No 2: 170-195, February 2000.

[Dahl, 1938] O.G.C. Dahl. *Electric Power Circuits. Vol.II: Power System Stability*. McGraw-Hill, 1938.

[Dercle, 1995] D. Dercle. "Dynamic Security Analysis. A New Method for Global Transient Stability Studies". In *Proc. of the IFAC Symp. on Control of Power Plants and Power Systems*, Cancun, Mexico, Vol.1: 261–266, December 1995.

[De Tuglie et al., 1999] E. De Tuglie, M. Dicorato, M. La Scala and P. Scarpellini. "Dynamic Security Preventive Control in a Deregulated Electricity Market". In *Proc. of the 13th PSCC'99*, Trondheim, Norway: 125-131, June 28th - July 2nd, 1999.

[Dy-Liacco, 1968] T.E. Dy-Liacco. *"Control of Power Systems via the Multi-Level Concept"*. Case Western Reserve University, Ohio, Rep. SRC-68-19, 1968.

[Dy-Liacco, 1970] T.E. Dy-Liacco. "The Emerging Concept of Security Control". In *Proc. of Symp. Power Systems*, Purdue University, Lafayette, Indiana, May 1970.

[Dy-Liacco, 1979] T.E. Dy-Liacco. "Security Functions in Power System Control Centers; The State-of-the-Art in Control Center Design". In *Proc. of IFAC Symp. on Computer Applications in Large Scale Power Systems*, New Delhi, India: 16–18, 1979.

[Dy-Liacco, 1996] T.E. Dy-Liacco. "On the Applicability of Automatic Learning to Power System Operation". *Revue E* (Special Issue on *Apprentissage Automatique; Applications aux Réseaux d'Energie Electrique*), 105-C: 19–22, December 1996.

[Dy-Liacco, 1999] T.E. Dy-Liacco. "Enabling Technologies for Operation and Real-Time Control in a Liberalized Environment". *EPRI - Second European Conference*, Vienna, November 2-4, 1999.

[Ejebe et al., 1997] G.C. Ejebe, C. Jing, J.G. Waight, V. Vittal, G. Pieper, F. Jamshidian, D. Sobajic, and P. Hirsch. "On-line Dynamic Security Assessment: Transient Energy Based Screening and Monitoring for Stability Limits". *IEEE Summer Meeting* (Panel Session on *Techniques for Stability Limit Search*), Berlin, Germany, 1997.

[EIA, 2000] Energy Information Administration (EIA) (On-line). Available at the website: http://eia.doe.gov/cneaf/electricity/page/restructure.html, 2000.

[El-Abiad and Nagappan, 1966] A.H. El-Abiad and K. Nagappan. "Transient Stability Region of Multimachine Power Systems". *IEEE Trans. on PAS*, PAS-85:169–179, 1966.

[EPRI, 1977] EPRI (Electric Power Research Institute) "Power system Dynamic Analysis, Phase I" Final EPRI Report No EPRI EL-0484, Projects 670-1, July 1977.

[EPRI, 1994] EPRI (Electric Power Research Institute) "Extended Transient Midterm Stability Program Version 3.1 User's manual" Final EPRI Report No EPRI TR-102004, EPRI, Projects 1208, 11-13 May, 1994.

[EPRI, 1995] EPRI (Electric Power Research Institute) "Standard Test Cases for Dynamic Security Assessment" Final EPRI Report No EPRI TR-105885, EPRI, Projects 3103-02, December, 1995.

[Ernst et al., 1998a] D. Ernst, A.L. Bettiol, D. Ruiz-Vega L. Wehenkel, and M. Pavella. "Compensation Schemes for Transient Stability Assessment and Control". In *Proc. of LESCOPE'98*, Halifax, Canada: 225–230, June 1998.

[Ernst et al., 1998b] D. Ernst, A.L. Bettiol, Y. Zhang L. Wehenkel, and M. Pavella. "Real-Time Transient Stability Emergency Control of the South-Southeast Brazilian System". In *Proc. of SEPOPE'98*, Salvador, Brazil, May 1998. (Invited paper, IP044; 9 pages).

[Ernst et al., 2000a] D. Ernst, D. Ruiz-Vega, and M. Pavella. "Preventive and Emergency Transient Stability Control". In *Proc. of SEPOPE'2000*, Curitiba, Brazil, May 23rd-28th, 2000. (Invited paper.)

[Ernst et al., 2000b] D. Ernst, D. Ruiz-Vega, M. Pavella, P. Hirsch, and D. Sobajic. "A Unified Approach to Transient Stability Contingency Filtering, Ranking and Assessment". Submitted.

[Ernst and Pavella, 2000] D. Ernst and M. Pavella. "Closed Loop Transient Stability Emergency Control". *IEEE PES Winter Meeting 2000, Singapore, (Panel Session: "On-Line Transient Stability Assessment and Control")*, January 2000.

[European Parliament, 1996] European Parliament. Directive 96/92 EC of the European Parliament and Council of 19th December, 1996.

[Fink and Carlsen, 1978] L.H. Fink and K. Carlsen. "Operating Under Stress and Strain". *IEEE Spectrum*, 15:43–58, 1978.

[Fonseca and Decker, 1985] L.G.S. Fonseca, and I.C. Decker. "Iterative Algorithm for Critical Energy Determination in Transient Stability of Power Systems". *IFAC Symp. on Planning and Operation of Electrical Energy Systems*, Rio de Janeiro, Brazil: 483-489, 1985.

[Fouad and Vittal, 1992] A.A. Fouad and V. Vittal. *Power System Transient Stability Analysis Using the Transient Energy Function Method.* Prentice-Hall, 1992.

[Gan et al., 1998] D. Gan, R.J. Thomas and R.D. Zimmerman. "A Transient Stability Constrained Optimal Power Flow". In *Proc. of the Bulk Power Systems Dynamics and Control IV - Restructuring".* Santorini, Greece, August 1998.

[Ghandhari et al., 2000] M. Ghandhari, G. Andersson, D. Ernst and M. Pavella. "A Control Strategy for Controllable Series Capacitor". Submitted.

[Gless, 1966] G.E. Gless. "Direct Method of Liapunov Applied to Transient Power System Stability". *IEEE Trans. on PAS*, PAS-85: 189–198, 1966.

[Granville, 1994] S. Granville. "Optimal Reactive Dispatch through Interior Point Methods". *IEEE Trans. on PWRS*, Vol.1, No 1: 136-146, 1994.

[Grujic et al., 1987] Lj.T. Grujic, A.A. Martynyuk, and M. Ribbens-Pavella. *Large Scale Systems Stability Under Structural and Singular Perturbations.* Springer-Verlag, 1987.

[Gorev, 1960] A.A. Gorev. *A Collection of Works on Transient Stability of Power Systems.* In *Second Lyapunov's Method and Its Application to Energy Systems.* Naouka Eds, Gosenergoirdat, 1966.

[Gupta and El-Abiad, 1976] C.L. Gupta and A.H. El-Abiad. "Determination of the Closest Unstable Equilibrium State for Liapunov Methods in Transient Stability Studies". *IEEE Trans. on PAS*, PAS-94:1699, 1976.

[Hiskens et al., 1999] I.A. Hiskens, M.A. Pai and T.B. Nguyen. "Dynamic Contingency Analysis Studies for Inter-Area Transfers". In *Proc. of 13th PSCC'99*, Trondheim, Norway: 345-350, June 28th - July 2nd, 1999.

[Hiskens and Pai, 2000] I.A. Hiskens and M.A. Pai. "Trajectory Sensitivity Analysis of Hybrid Systems". *IEEE Trans. on Circuits and Systems*, Part I, Vol. 47, No 2: 204–220, February 2000.

[Houben et al., 1997] I. Houben, L. Wehenkel and M. Pavella. "Hybrid Adaptive Nearest Neighbor Approaches to Dynamic Security Assessment". In *Proc. of IFAC/CIGRE Symp. on Control of Power Systems and Power Plants*, Beijing, China, August 1997.

[Hunt and Shuttleworth, 1996] S. Hunt and G. Shuttleworth. *Competition and Choice in Electricity*, John Wiley and Sons, 1996.

[IEEE, 1968] IEEE Committee Report. "Computer Representation of Excitation Control Systems", *IEEE Trans. on PAS*, PAS-87, No 6: 1460-1464, June 1968.

[IEEE, 1996] IEEE Power Engineering Society. "Glossary of Terms Concerning Electric Power Transmission System Access and Wheeling", *IEEE PES*, PES-96,TP 110-0, 1996.

[Ilic et al., 1998] M. Ilic, F. Galiana and L. Fink. *Power System Restructuring: Engineering and Economics*, Kluwer Academic, 1998.

[Kakimoto et al., 1980] N.Y. Kakimoto, Y. Ohsawa, and M. Hayashi. "Transient Stability Analysis of Multimachine Power Systems With Field Flux Decays via Lyapunov's Direct Method". *IEEE Trans. on PAS*, PAS-99: 1819-1827, 1980.

[Kalman, 1963] R.E. Kalman. "Lyapunov Functions for the Problem of Luré in Automatic Control". *Proc. Nat. Acad. of Sciences* (USA), 49: 201–206, 1963.

[Kimbark, 1948] E.W. Kimbark. *"Power System Stability"*. John Wiley & Sons, 1948.

[Koizumi et al., 1975] K. Koizumi, O. Saito, K. Masegi, and M. Udo. "Fast Transient Stability Study Using Pattern Recognition". *5th PSCC'75, Cambridge*, 1975.

[Kokotovic and Rutman, 1965] P.V. Kokotovic and R. Rutman. "Sensitivity of Automatic Control Systems" (Survey). *Automatic and Remote Control*, 26: 727–748, 1965.

[Kundur, 1994] P. Kundur. *Power System Stability and Control*. McGraw-Hill, 1994.

[Kundur and Morisson, 1997a] P. Kundur and G.K. Morison. "A Review of Definitions and Classification of Stability Problems in Today's Power Systems". *IEEE-PES Winter Meeting (Panel Session on Stability Terms and Definitions*, New York, USA, February 2–6, 1997.

[Kundur and Morisson, 1997b] P. Kundur and G.K. Morison. "Techniques for Emergency Control of Power Systems and Their Implementation". In *Proc. IFAC-CIGRE Symp. on Control of Power Plants and Power Systems* Beijing, China: 679–684, 1997.

[Kundur and Morisson, 1998] P. Kundur and G.K. Morison. "Power System Control: Requirements and Trends in the New Utility Environment". In

Proc. of the Bulk Power Systems Dynamics and Control IV - Restructuring, Santorini, Greece: 257-263, August 1998.

[Laufenberg and Pai, 1997] M.J. Laufenberg and M.A. Pai. "A New Approach to Dynamic Security Assessment Using Trajectory Sensitivities". In *Proc. of PICA'97*, Columbus, OH, USA, May 11-16, 1997.

[Lyapunov, 1907] A.M. Lyapunov. "Problème Général de la Stabilité du Mouvement". French translation in 1907 of russian edition, Commun. Soc. Math. Kharkow, 1893. Reprinted in *Annals of Mathematical Studies*, 17, Princeton University Press, 1949.

[Magnusson, 1947] P.C. Magnusson. "Transient Energy Method of Calculating Stability". *AIEE Trans.*, 66, 1947.

[Maria et al., 1990] G.A. Maria, C. Tang, and J. Kim. "Hybrid Transient Stability Analysis". *IEEE Trans. on PWRS*, Vol. 5, No 2: 384–393, May 1990.

[MATLAB, 1999] "MATLAB program, Version 5.2". *The MathWorks Inc.*, 1999.

[Mello et al., 1997] J.C.O. Mello, A.C.G. Melo and S. Granville. "Simultaneous Transfer Capability Assessment by Combining Interior Point Methods and Monte Carlo Simulation". *IEEE Trans. on PWRS*, Vol. 12, No 2: 736-742, 1997.

[Merlin et al., 1997] A. Merlin, P. Bornard, C. Gallaire, D. Haag, J. Tesseron and P. Virleux. "Le Rôle du Gestionnaire du Réseau de Transport dans le Système Electrique Français". In *Proc. of the CIGRE Tours Symposium*, Tours, France, June 8-10, 1997.

[Moore and Anderson, 1968] J.B. Moore, and B.D.O. Anderson. "A Generalization of Popov Criterion". *J. Franklin Institute*, 285: 488–492, 1968.

[NERC, 1997] North American Electric Reliability Council (NERC). "Available Transfer Capability Definitions and Determination" (On-line). Available at the website: http://www.nerc.com.

[Pai and Narayana, 1975] M.A. Pai, C.L.Narayana. "Stability of Large Scale Power systems" In *Proc 6th IFAC World Congress*, Boston 1975.

[Pai, 1981] M.A. Pai, Editor. *Power System Stability Analysis by the Direct Method of Lyapunov*. North-Holland, 1981.

[Pai, 1989] M.A. Pai, Editor. *Energy Function Analysis for Power System Stability*. Kluwer Academic, 1989.

[Park and Bancker, 1929] R.H. Park and E.H. Bancker. "System Stability as a Design Problem". *AIEE Trans.*, 48:170–194, 1929.

[Pavella and Murthy, 1993] M. Pavella and P.G. Murthy. *Transient Stability of Power Systems, Theory and Practice*. John Wiley, 1993.

[Pavella et al., 1997] M. Pavella, L. Wehenkel, A. Bettiol, and D. Ernst. "An Approach to Real-Time Transient Stability Assessment and Control". *IEEE Summer Meeting, Berlin, Germany (Panel Session: "Techniques for Stability Limit Search")*, TP-138-0, 1997.

[Pavella, 1998] M. Pavella. "Generalized One-Machine Equivalents in Transient Stability Studies". *PES Letters, IEEE Power Engineering Review*, 18(1):50–52, January 1998.

[Pavella and Wehenkel, 1998] M. Pavella and L. Wehenkel. "Electric Power System Dynamic Security Assessment". *Revue Internationale de Génie Electrique*, Vol.1, No.1: 667–698, 1998.

[Pavella et al., 1999a] M. Pavella, D. Ernst, and D. Ruiz-Vega. "A General Framework for Transient Stability Assessment and Control". *PES Letters, IEEE Power Engineering Review*, Vol. 19, No 10: 1699, October 1999.

[Pavella et al., 1999b] M. Pavella, D. Ruiz-Vega, J. Giri, and R. Avila-Rosales. "An Integrated Scheme for On-Line Static and Transient Stability Constrained ATC Calculations". *IEEE Summer PES Meeting* (Panel Session on *On-Line DSA Projects for Reliability Management and ATC Computation*), Edmonton, Canada, 1999.

[Phadke, 1993] A.G. Phadke. "Synchronized Phasor Measurements in Power Systems". *IEEE Computer Applications in Power*, Vol. 6, No 2: 10–15, April 1993.

[Prabhakara and El-Abiad, 1975] F.S. Prabhakara and A.H. El-Abiad. "A Simplified Determination of Transient Stability Regions for Lyapunov Methods". *IEEE Trans. on PAS*, PAS-94:672–689, 1975.

[Prabhakara and Heydt, 1987] F.S. Prabhakara, and G.T. Heydt. "Review of Pattern Recognition Methods for Rapid Analysis of Transient Stability Assessment". *IEEE PES Winter Meeting*, New York, Paper 87THO 169-3 PWR, 1987.

[Putilova and Tagirov, 1970] A.T. Putilova and M.A. Tagirov. "Liapunov Functions for Reciprocal Equations of Motion of Synchronous Machines". *Reports of the Siberian Institute of Scientific Power Researchers*, 17: Stability of Power Systems:672–689, 1970.

[Rahimi and Schaffer, 1987] F.A. Rahimi, and G. Schaffer. "Power System Transient Stability Indexes for On-Line Analysis of 'Worst Case Dynamic Contingencies". *IEEE Trans. on PWRS*, PWRS-2: 660–668, 1987.

[Rahimi, 1990] F.A. Rahimi. "Generalized Equal-Area Criterion: A Method for On-Line Transient Stability Analysis". In *Proc. of IEEE Int. Conf. on Systems, Man, and Cybernetics*, Los Angeles, California, 684–688, November 1990.

[Ribbens-Pavella and Evans, 1981] M. Ribbens-Pavella and F.J. Evans. "Direct Methods for Studying Dynamics of Large-Scale Electric Power Systems - A Survey". *Automatica*, 21: 1–21, 1981.

[Ribbens-Pavella et al., 1981] M. Ribbens-Pavella, P.G. Murthy, and J.L. Horward. "An Acceleration Approach to Practical Stability Domain Estimation in Power Systems". In *Proc. of the 20th IEEE Conf. on Decision and Control*, San Diego, 1: 471–477, 1981.

[Ribbens-Pavella et al., 1982a] M. Ribbens-Pavella, P.G. Murthy, J.L. Horward and J.L. Carpentier. "Transient Stability Index for On-Line Stability Assessment and Contingency Evaluation". *Journal of EPES*, Vol. 4, No 2, April 1982.

[Ribbens-Pavella et al., 1982b] M. Ribbens-Pavella, P.G. Murthy, J.L. Horward and J.L. Carpentier. "On-Line Transient Stability Assessment and Contingency Analysis". *CIGRE, 1982 Session*, Paper No 32-19, Paris, France, September 1-9, 1982.

[Rovnyak et al., 1997] S.M. Rovnyak, C.W. Taylor and J.S. Thorp. "Performance Index and Classifier Approaches to Real-Time, Discrete-Event Control". *Control Engineering Practice*, Vol. 5, No 1: 91–99, 1997.

[Rozenvasser, 1960] E.N. Rozenvasser. "On the Construction of a Lyapunov Function for a Class of Nonlinear Systems" (in Russian). *Izv. Akad. Nauk SSSR, Otd. Tekh. Nauk.*, No.2.

[Ruiz-Vega et al., 1998] D. Ruiz-Vega, A.L. Bettiol, D. Ernst, L. Wehenkel, and M. Pavella. "Transient Stability-Constrained Generation Rescheduling". In *Bulk Power System Dynamics and Control IV - Restructuring*, Santorini, Greece: 105–115, August 1998.

[Ruiz-Vega et al., 2000a] D. Ruiz-Vega, D. Ernst, C. Machado Ferreira, M. Pavella, P. Hirsch, and D. Sobajic. "A Contingency Filtering, Ranking and Assessment Technique for On-line Transient Stability Studies". In *Proc. of the DRPT2000 Conf.*, London, UK: 459-464, April 2000.

[Ruiz-Vega et al., 2000b] D. Ruiz-Vega, D. Ernst, and M. Pavella. "Integrated Schemes for On-line Transient Stability Constrained ATC Calculations". (Invited paper) *"VII SEPOPE' 2000*, Curitiba, Brazil", May 23rd-28th, 2000.

[Sauer and Pai, 1998] W.P. Sauer and M.A. Pai. *Power System Dynamics and Stability*, Prentice Hall, Englewood Cliffs, NJ,1998.

[Sheblé, 1999] G.B. Shebblé. "Computational Auction Mechanisms for Restructured Industry Orperation" Kluwer Academic Publishers, 1999.

[Shirmohammadi et al., 1998] D. Shirmohammadi, B. Wollenberg, A. Vojdani, P. Sndrin, M. Pereira, F.Rahimi, T. Schneider and B. Stott. "Transmission Dispatch and Congestion Management in the Emerging Energy Management Structures". *IEEE Trans. on PWRS*, Vol. PWRS-13, No 4, November 1998.

[Skilling and Yamakawa, 1940] H.H. Skilling and M.H. Yamakawa. "A Graphical Solution of Transient Stability". *Electrical Engineering*, Vol 59: 462–465, 1940.

[Sterling et al., 1991] J. Sterling, M.A. Pai and P.W. Sauer. "A Methodology to Secure and Optimal Operation of a Power System for Dynamic Contingencies". *Journal of Electric Machines and Power Systems*, Vol 19, No 5, September/October 1991.

[Russell and Smart, 1997] T.D. Russell and A.L. Smart. "The System Operator in an Open Electricity Market Meeting the Challenge". In *Proc. of the CIGRE Tours Symposium*, Tours, France, June 8-10, 1997.

[FERC, 1996] Federal Energy Regulation Commission (FERC) Order 888. Final Rule, "Promoting Wholesale Competition through Open Access Non-Discriminatory Transmission Services by Public Utilities; Recovery of Stranded Cost by Public Utilities and Transporting Utilities", Docket Nos. RM98-8-00 and RM 94-7-001, April 1996.

[Valette et al., 1987] A. Valette, F. Lafrance, S. Lefebvre, and L. Radakovitz. "ST600 Programme de Stabilité: Manuel d'Utilisation Version 701". *Technical Report*, Hydro-Québec, Vice-Présidence Technologie et IREQ, 1987. *CIGRE Session, Paris, France*, September 1999.

[Wehenkel and Pavella, 1996] L. Wehenkel, and M. Pavella. "Why and Which Automatic Learning Approaches to Power Systems Security Assessment". In *Proc. of CESA'96, IMACS./IEEE SMC Multiconference on Computational Engineering in Systems Applications*, Lille, France: 1072–1077, 1996.

[Wehenkel et al., 1997] L. Wehenkel, C. Lebrevelec, M. Trotignon, and J. Batut. "A Probabilistic Approach to the Design of Power System Protection Schemes Against Blackouts". In *Proc. of IFAC/CIGRE Symp. on Control of Power Systems and Power Plants*, Beijing, China: 506–511, 1997.

[Wehenkel, 1998] L. Wehenkel. *Automatic Learning Techniques in Power Systems*. Kluwer Academic Publishers, 1998.

[Wehenkel, 1999] L. Wehenkel. "Emergency Control and Its Strategies". In *Proc. of the 13th PSCC'99*, Trondheim, Norway: 35–48, June 28 - July 2, 1999.

[Willems, 1969] J.L. Willems. "The Computation of Finite Stability Regions by Means of Open Liapunov Surfaces". *Int. Journal Control*, 10: 537–544, 1969.

[Xue et al., 1986] Y. Xue, Th. Van Cutsem, and M. Ribbens-Pavella. "A New Decomposition Method and Direct Criterion for Transient Stability Assessment of Large-Scale Electric Power Systems". *IMACS/IFAC Symp. on Modeling and Simulation*, Lille, France: 183–186, 1986.

[Xue et al., 1988] Y. Xue, Th. Van Cutsem, and M. Ribbens-Pavella. "A Simple Direct Method for Fast Transient Stability Assessment of Large Power Systems". *IEEE Trans. on PWRS*, PWRS-3: 400–412, 1988.

[Xue, 1988] Y. Xue. *A New Method for Transient Stability Assessment and Preventive Control of Power Systems*. PhD Thesis, University of Liège, Belgium, December 1988.

[Xue et al., 1992] Y. Xue, L. Wehenkel, R. Belhomme, P. Rousseaux, M. Pavella, E. Euxibie, B. Heilbronn, and J.F. Lesigne. "Extended Equal-Area Criterion Revisited". *IEEE Trans. on PWRS*, PWRS-7: 1010-1022, 1992.

[Xue and Pavella, 1993] Y. Xue, and M. Pavella. "Critical Cluster Identification in Transient Stability Studies". *IEE Proc.*, 148 - PtC, 1993.

[Xue et al., 1993] Y. Xue, P. Rousseaux, L. Wehenkel, Z. Gao, M. Pavella, M. Trotignon, A. Duchamp, and B. Heilbronn. "Dynamic Extended Equal-Area Criterion. Part II: Recent Extensions". *Athens Power Tech, NTUA-IEEE/PES Joint Int. Power Conf.*, September 1993.

[Xue, 1996] Y. Xue "Integrated Extended Equal Area Criterion - Theory and Application" In *Proc. of VI SEPOPE'96*, Brazil, 1996.

[Xue et al., 1997] Y. Xue, Y. Yu, J. Li, Z. Gao, C. Ding, F. Xue, L. Wang, G.K. Morison, and P. Kundur. "A New Tool for Dynamic Security Assessment of Power Systems". *IFAC/CIGRE Symposium on Control of Power Systems and Power Plants*, Beijing, China, August 18-21, 1997.

[Zhang, 1993] Y. Zhang. "A Mixed DEEAC Approach". Private communication, March 1993.

[Zhang, 1995] Y. Zhang. *Hybrid Extended Equal-Area Criterion: A General Method for Transient Stability Assessment of Multimachine Power Systems*. PhD Thesis, University of Liège, February 1995.

[Zhang et al., 1995] Y. Zhang, P. Rousseaux, L. Wehenkel, M. Pavella, Y. Xue, B. Meyer, and M. Trotignon. "Hybrid Extended Equal-Area Criterion for Fast Transient Stability Assessment with Detailed System Models". *IFAC Symp. on Control of Power Plants and Power Systems (SIPOWER)*, Cancun, Mexico, December 1995.

[Zhang et al., 1996] Y. Zhang, L. Wehenkel, P. Rousseaux, and M. Pavella. "First- and Multi-Swing Transient Stability Limits of a Longitudinal System Using the SIME Method". In *Proc. of MELECON'96*, Bari, Italy: 809–815, May 1996.

[Zhang, 1996] Y. Zhang. "Compensation Schemes for Approximating Critical Clearing Times". Private communication, December 1996.

[Zhang et al., 1997a] Y. Zhang, L. Wehenkel, P. Rousseaux, and M. Pavella. "SIME : A Hybrid Approach to Fast Transient Stability Assessment and Contingency Selection". *Journal of EPES*, Vol 19, No 3: 195–208, 1997.

[Zhang et al., 1997b] Y. Zhang, L. Wehenkel, and M. Pavella. "A Method for Real-Time Transient Stability Emergency Control". In *Proc. CPSPP'97, IFAC/CIGRE Symp. on Control of Power Systems and Power Plants*, Beijing, China: 673–678, August 1997.

[Zhang et al., 1998] Y. Zhang, L. Wehenkel, and M. Pavella. "SIME : A Comprehensive Approach to Fast Transient Stability Assessment". In *Trans. of IEE Japan*, Vol 118-B, No 2: 127–132, 1998.

Index

analysis
    predictive, 176
    preventive, 93
ANN, 27, 203
automatic learning, **23**, 189, 201
    ANN, 27, 203
    decision tree, 25, 203
    kNN, 28, 205
available transfer capability, **11**, 166
AVR, 58, 72, 215

backswing, **60**, 160, 179, 191
    instability, 43, 59, 157

clearing time
    critical clearing time, 96
    initial conditions, 98
    linear behaviour, 81
    range of, 50
compensation scheme, 83
    for critical clearing times, 83
    for power limits, 87
computing performances, 77, 95, 117, 149, 155, 163
congestion management, **10**, 139, 168
contingency
    assessment, 115
    filtering, 115, 154
    first-swing stable, 116
    harmful, 116
    harmless, 116
    potentially harmful, 116
    ranking, 115
control, 62, 64
    closed-loop emergency, 198
    open-loop emergency, 197
    preventive, 93, 128
    preventive vs emergency, 193, 197
    time, 175
control center
    TSA&C, 165
critical clearing time, 96
    first-swing, 59, 191
    linearized approximation, 81
    multiswing, 59, 191
critical machines, **38**, 40, 62
    generation reallocation, 106
    identification, 38

decision tree, 25, 203
direct methods, **17**
DTS, 166
    optimal power flow, 139
    TSA&C, 169

EAC, 34, 41, 207
EEAC, 34, 36
emergency control, 172, 179
    closed-loop, 172
    open-loop, 172
EMS, 12, 31, 139, **165**, 167, 191

FACTS, 14, 179
FASTEST, 35
FILTRA, 115, 116, 140
    classification ability, 116
    computation efficiency, 116
    efficacy, 116
    false alarm, 125
    filtering block, 118
    main properties, 121
    ranking and assessment block, 118
    reliability, 116
first-swing, 42, 43, 191
    CCT, 59, 98
    instability, 43, 96, 145
    time to instability, 59, 191

generation, 137

236   TRANSIENT STABILITY OF POWER SYSTEMS

reallocation, 89
    principle of, 128, 132
shedding, 180
shifting, 137

instability phenomena
    backswing, 60
    characterization, 191
    first-swing, 42, 145
    multiswing, 42, 191
    upswing, 191
inter-area mode
    instability constraints, 157
    oscillation, 134
ISO, 10, 166
    on-line TSA&C, 167

Lyapunov, **18**
    direct method, 18
    vector function, 35

machine
    angle, 173
    critical, 37, 38, 40, 149
        identification, 38, 180
    inertia, 41, 73
    non-critical, 40, 170
    power limit, 88
    speed, 173
maximum allowable transfer, 141, 144, 164
measurements
    real-time, 36, 173, 184
models, **14**
    FACTS, 14
    load, 14
    machine, 14
    transient stability, 14
multiswing, **42**, 59
    CCT, 59, 98
    instability, 43, 96, 157
    time to instability, 59, 191

non-critical machines, 37, 40

OMIB, 33, 39, 56
    angle, 42
        equilibrium, 44
            return, 42, 44
            stable, 42
            unstable, 42, 43
    critical, 37
        identification, 43
    electrical power, 42
    generalized, 34
    identification, 38
    inertia, 41
    mechanical power, 42

parameters, 40
phase plane, 56
power limit, 87
power linear behaviour, 81
power-angle curves, 56
speed, 42
stable trajectory, 43, 44
swing curves, 56
time-invariant, 34
time-varying, 34
transformation, 33
unstable trajectory, 43
Optimal Power Flow, 12, **142**, 190
oscillation, 159, 192
    backward, 61
    inter-area mode, 134, 157
    plant mode, 145
    upward, 160

phase plane, 103
plant mode
    instability, 145
    instability constraints, 145
power
    flow, 152
    losses, 152, 164
power limit, 102
    computation, 102
    definition, 102
    linearized approximation, 81
predictive SIME, 175
preventive control, 172

restructured electric industry, 9, 165
rotor, 35
    angle, 2, 40, 173
        OMIB, 40
    speed, 40, 173
        OMIB, 40

security, **3**
    dynamic, 3, 139
    static, 3, 141
sensitivity, **68**
    analysis, 69
    function, 71
    synthetic function, 71
sensitivity analysis, 62
SIME, 33, 37
    accuracy, 100, 125
    as a reduction technique, 61
    coupling, 37, 191
    emergency, 63, 192
    fundamentals, 35
    line analysis, 198
    predictive, 64, 175
    preventive, 63, 190
    principle, 37

Index

sensitivity
    analysis, 69
    transformation, 33
    typical representations, 55
software
    integrated, 139, 143
    TSA&C, 140
stability limit
    approximate assessment, 112
    computation of, 96, 102
stability margin, 45, 62
    analytical expressions, **62**
    existence of, 49
    extrapolation, 96, 105
    interpolation, 96, 105
    linearized approximation, 81
    normalized, 55
    range of, 49
    stable, 46
    substitute for, 52
    triangle, 47
    unstable, 45
    variation with clearing time, 81
    WLS, 48
stabilization
    multiple contingencies, 132
    simultaneous, 95

single contingency, 128
stable angle, 42
state variable, 14, 58, 77, 219
stopping criteria, 114

time to instability, 43, 98, 191
time-domain, **15**, 203
    approaches, 15
    pros and cons, 16
time-invariant, 34
time-varying parameters
    OMIB, 39
transient stability, **1, 4**
    assessment and control, 63, 128
    definition, 1
    models, 13
    predictive assessment, 64
    preventive assessment, 63

unbundling, 9
unstable angle, 42

validity test
    prediction, 173

weighted least-squares, 48, 177